Environmental Toxin Series 1

Editors-in-Chief: S. Safe and O. Hutzinger

Environmental Toxin Series

Editors-in-Chief: S. Safe and O. Hutzinger

In preparation: Volume 2   Cadmium
M. Stoeppler and M. Piscator (Eds.)

Contents: I. Toxicity, Carcinogenicity, Animal Experiments
II. Epidemiology
III. Cadmium in the Environment
IV. Methodology and Quality Assessment

S. Safe (Ed.)

Polychlorinated Biphenyls (PCBs): Mammalian and Environmental Toxicology

With Contributions by
R. L. Dedrick L. G. Hansen M. A. Hayes
R. J. Lutz M. Mullin A. Parkinson L. Safe S. Safe
R. G. Schnellmann I. G. Sipes

With 33 Figures and 35 Tables

Springer-Verlag
Berlin Heidelberg New York
London Paris Tokyo

Volume Editor

Prof. Dr. Stephen Safe

Texas A&M University, College of Veterinary Medicine,
Department of Veterinary Physiology and Pharmacology,
College Station, TX 77843-4466, USA

ISBN 3-540-15550-3 Springer-Verlag Berlin Heidelberg New York
ISBN 0-387-15550-3 Springer-Verlag New York Berlin Heidelberg

Library of Congress Cataloging in Publication Data. Polychlorinated biphenyls (PCBs).
(Environmental toxin series; v. 1)
1. Polychlorinated biphenyls – Toxicology. 2. Polychlorinated biphenyls – Environmental
aspects. 3. Polychlorinated biphenyls – Physiological effect. I. Safe, S. II. Dedrick,
Robert L. III. Series. RA1242.P7P65 1987 615.9′512 87-16437

Typesetting, printing and binding: Brühlsche Universitätsdruckerei, Giessen
2154/3140-543210

Editorial

The concern about environmental toxins is ever increasing, as is the need for sound scientific information. The Environmental Toxin Series is dedicated to the publication of comprehensive reviews and monographs on compounds or classes of chemicals which are of importance in environmental toxicology. The series is designed to serve as a background of information for scientific investigation as well as risk analysis and political decision making. The main aim of the series is to describe in as complete a way as possible all potentially hazardous chemicals from the point of view of chemistry, ecology, toxicology, risk analysis and regulatory implications. From time to time conference proceedings on important and urgent topics will be included in the series. We thank the members of the editorial board for their enthusiastic support.

S. Safe and O. Hutzinger

Contents

Editorial Board

Polychlorinated Biphenyls:
Environmental Occurrence and Analysis

S. Safe [1], L. Safe [1] and M. Mullin [1]

Polychlorinated biphenyls (PCBs) are complex mixtures which have been identified in every component of the global ecosystem. This chapter discusses the distribution of PCBs in the environment and points out recent analytical advances which now permit high resolution congener-specific analysis of PCBs in diverse analytes using high resolution capillary gas chromatographic techniques. The biologic and toxic effects of PCBs are structure-dependent and the adverse environmental and human health impacts of the different mixtures of PCBs are related to the individual components of these mixtures and their interactions. The chapter points out that high resolution PCB analysis will now permit the unequivocal identification and quantitation of the individual PCBs in environmental samples and this data can be used for more accurate risk assessment

1 Introduction

Polychlorinated biphenyls (PCBs) are commercial products which are prepared industrially by the chlorination of biphenyl and the commercial preparations are graded and marketed according to their chlorine content. The lower chlorinated materials (e.g. Aroclor 1221; containing 21% by weight of chlorine) are moderately viscous liquids whereas the more highly chlorinated products, such as Aroclor 1260 (60% by weight of chlorine) are solids. Commercial PCBs have been

---

[1] Texas A&M University, College of Veterinary Medicine, Dep. of Veterinary Physiology and Pharmacology, College Station, TX 77843-4
and
U.S. Environmental Protection Agency, Large Lakes Research Station, Grosse Ile, MI 48138, USA

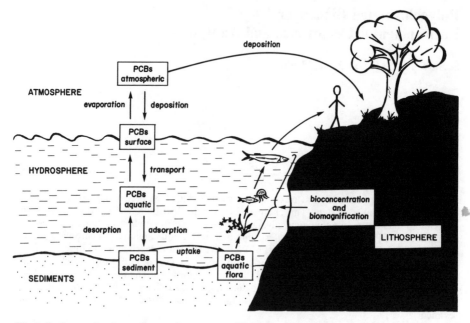

Fig. 1. Pathways for the cycling and transport of PCBs in the environment (42)

produced in several countries including the United States, France, Japan, Czechoslovakia and Russia, and have been marketed and used worldwide.

PCBs have been widely used in industry as heat transfer fluids, hydraulic fluids, solvent extenders, flame retardants, organic diluents, and dielectric fluids (1). The unusual industrial versatility of PCBs is directly related to their physical properties which include resistance to acids and bases, compatibility with organic materials, resistance to oxidation and reduction, excellent electrical insulating properties, thermal stability, and nonflammability. The widespread use of PCBs coupled with improper disposal practices has led to significant environmental contamination by commercial PCB formulations. Some estimates suggest that up to one-third of the total United States production of PCBs (approximately 1.4×10^9 lbs) has entered the environment (2). Not surprisingly, PCBs are highly stable in the environment and are readily transported from localized or regional sites of contamination throughout the global ecosystem (3, 4). However, PCB residues have also been reported in regions of no industrial activity, for example residues have been detected in snow deposits in the Antarctic (6, 7). The lipophilic nature and persistence of PCBs also contributes to their high bioaccumulation potential and their biomagnification in higher trophic levels of the food chain. PCB residues are routinely detected in fish, wildlife, and human adipose tissue, blood and milk (8–30). Figure 1 shows a schematic diagram which illustrates the environmental fate and transport of PCBs from different matrices. Evaporation and deposition are clearly the most important routes for the long range transport of these industrial pollutants and biomagnification of these highly lipophilic contaminants contributes to their high levels in food chain animals and humans.

2 PCB Analysis: Progress and Problems

There are 209 possible PCB isomers and congeners and the chlorination of biphenyl results in the formation of a complex mixture of individual components. Not surprisingly, all the commercial PCB formulations and environmental extracts are mixtures. The analysis of commercial PCB mixtures by packed column and glass capillary column gas chromatography has been reported by several groups (31–42) and Figures 2–4 illustrate the high resolution gas chromatographic analysis of Aroclor 1016, a mixture of Aroclors 1016, 1254 and 1260 (5:4:2) and Aroclor 1260 using a glass capillary column equipped with electron capture detection (42 and M. Mullin, unpublished results). At least 80 individual PCB peaks were obeserved in Aroclor 1260 and over 100 PCBs were identified in the mixture of the 3 commercial products. In contrast, packed column gas chromatograms of these mixtures are much less complex due to multiple peak overlaps and the low resolving power of these columns (43–44). Most analytical schemes utilize gas chromatography for separation of the PCB components in cleaned up environmental extracts and electron capture (EC) detection for quantitation of individual or "total" PCBs. GC-mass spectrometry can be employed to confirm the molecular weights of individual PCBs and identify the presence of coextractives and other environmental contaminants such as DDT, DDE, lindane and other organochlorine contaminants which are routinely present in environmental samples

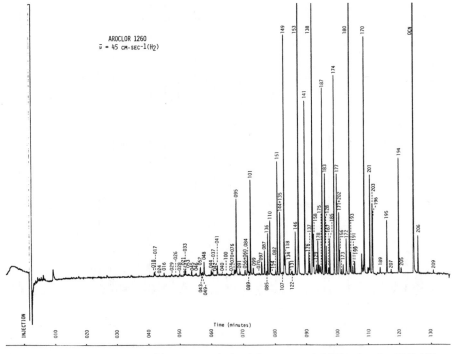

Fig. 2. High resolution glass capillary GC analysis of the commercial PCB, Aroclor 1260 (42)

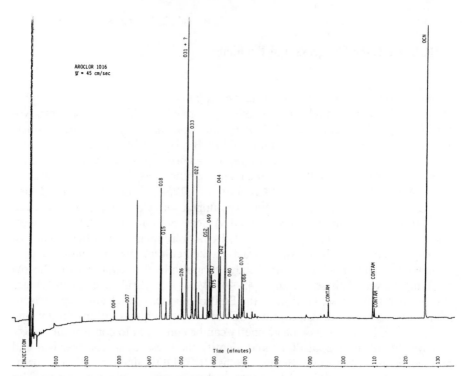

Fig. 3. High resolution glass capillary GC analysis of the commercial PCB, Aroclor 1016

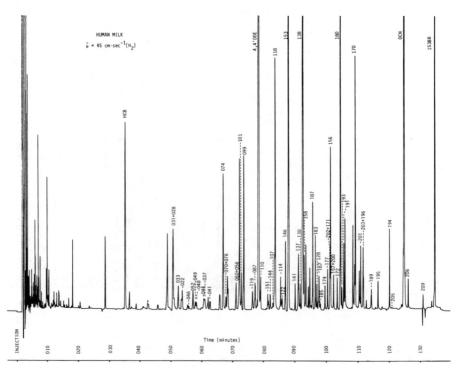

Fig. 4. High resolution glass capillary GC analysis of a mixture of Aroclors 1016, 1254 and 1260 (5:4:2)

(45). Successful environmental monitoring for PCBs is also dependent on the development of several matrix-specific clean up procedures which maximize the recovery of the PCBs and minimize the presence of unwanted coextractives.

Most analytical schemes for PCBs use the various commercial PCB preparations as quantitative reference standards (43, 44). The PCB concentrations are estimated by comparing the relative intensities of several diagnostic peaks observed in the commercial reference standards with their intensities in the analyte. More recent studies have utilized more specific patterns or isomer groups for recognition and "typing" of PCBs in diverse matrices (46–49). However, the accuracy in determining PCB levels is highly variable and matrix dependent; for example, PCBs which occur in many waste industrial fluids or in (retro-filled) transformers have not been significantly degraded and usually resemble a specific commercial mixture and therefore comparative low resolution packed column GC analysis can yield accurate quantitative results. In contrast, gas chromatographic analysis of PCBs in extracts from diverse environmental matrices clearly indicates that these mixtures can be strikingly different from the commercial PCB products which are used to quantitate total PCB levels. These differences in composition reflect the major differences in the physical properties (e.g., water solubility and volatility) and biodegradability of the individual PCBs present in the commercial mixtures. Figure 1 illustrates some of the processes which may change the composition of a commercial PCB preparation that has been introduced into an

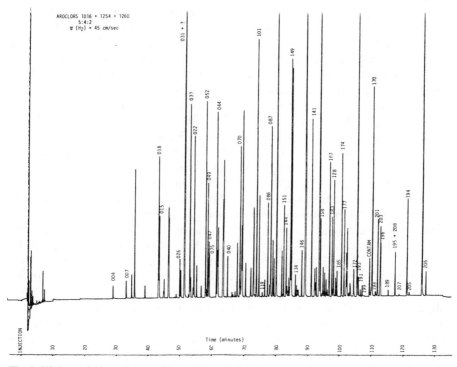

Fig. 5. High resolution glass capillary GC analysis of a composite human milk sample extract from the State of Michigan (42)

aquatic or marine environment; these include physical partitioning between the water-sediment and water-air interfaces, sediment desorption processes, and bio-magnification and bioconcentration with aquatic forms of life and the food chain. These processes differentially effect individual PCB congeners and therefore, it is not surprising that the composition of PCB extracts from these diverse environmental matrices can vary widely and often do not resemble any commercial mixture (31, 37, 42, 47). Quantitative analysis of these mixtures is usually determined by pattern or peak matching methods using artificially reconstituted mixtures of different commercial PCB formulations (e.g. Figure 4). At best, these results are only semiquantitative estimates of the "total PCB" levels in these environmental samples. Figure 5 illustrates the high resolution glass capillary gas chromatogram of a PCB-containing extract of a composite human milk sample (42). It is evident that the composition of this extract is markedly different from the composition of the commercial PCB mixtures.

3 High Resolution PCB Analysis

High-resolution glass capillary GC analysis can provide a solution to some of the analytical problems noted above (31–42). The high resolving power of coated silica or quartz capillary columns offers a method that can separate the PCBs present in most samples; the identities of the individual peaks must then be determined by using synthetic standards and/or by retention index addition methods (52). This later technique predicts the relative retention times (RRTs) of specific PCBs and has been used to assign the structures of the individual PCB congeners present in diverse analytes. This method relies on the RRT values that have been determined for the limited number of available synthetic PCB standards. However, accurate quantitation of the individual PCB components in a mixture can only be accomplished by comparing the observed relative retention time (RRT) and peak height (or area) data for a PCB-containing extract and the RRT and molar (or weight) response factors for all the PCB standards.

The unambiguous synthesis of all 209 PCB isomers and congeners has been reported (36). Most of these compounds were synthesized via diazo coupling of commercially available or synthetic chlorinated anilines and chlorinated benzenes to give either pure coupling products or defined mixtures. A total of 142 pure PCB congeners were synthesized and characterized by mass and proton magnetic resonance spectroscopy; 67 PCBs were synthesized as defined mixtures of 2–3 compounds. Table 1 summarizes the gas chromatographic properties of the 209 PCBs using a Varian model 3700 gas chromatograph equipped with a ^{63}Ni electron capture detector. The relative retention times (RRTs) and response factors were determined using a 50 m fused silica capillary column coated with SE-54. The RRT values for the PCBs were highly dependent on the degree of chlorination and chlorine substitution patterns. For example; the RRT values for the (i) monochlorophenyl-substituted PCBs increased in the order 2- < 3- < -4, (ii) the dichlorophenyl-substituted PCBs increased in the order 2,6- < 2,5- < 2,4- < 2,3- < 3,5- < 3,4-, (iii) the trichlorophenyl-substituted PCBs increased in the order 2,4,6- < 2,3,6-

<2,3,5-<2,4,5-<2,3,4-<3,4,5-, (iv) the tetrachlorobiphenyl-substituted PCBs increased in the order 2,3,5,6-<2,3,4,6-<2,3,4,5-. It was also apparent that the retention times of the PCB homologs within each series increased with decreasing ortho-chloro substituents.

Mullin and coworkers (36) successfully separated 187 of the 209 PCB congeners by capillary GC; however, 11 pairs of compounds, namely, 94/61, 70/76, 95/80, 60/56, 145/81, 144/135, 140/139, 133/122, 163/160, 202/171 and 203/196 exhibited similar retention times using the SE-54 coated glass capillary column. However, it was noted that some of these pairs of PCBs (e.g., 203/196) can be resolved chromatographically despite their comparable RRT values. In addition, it was also evident that some of these compounds were not present in the commercial PCBs.

Table 1. Relative retention times and response factors for 209 PCB congeners

Isomer[P] No.	Relative Retention time	Relative response factor	Isomer[P] No.	Relative Retention time	Relative response factor
0	0.0997	0.0251	32	0.3636	0.278
1	0.1544	0.0393	33	0.4163	0.447
2	0.1937	0.04[a]	34	0.3782	0.6092
3	0.1975	0.0193	35	0.4738	0.3746
4	0.2245	0.0374	36	0.4375	0.2948
5	0.2785	0.119	37	0.4858	0.58
6	0.2709	0.38	38	0.4593	0.4698
7	0.2566	0.69	39	0.4488	0.347
8	0.2783	0.206	40	0.5102	0.722
9	0.257	0.388	41	0.499	0.5469
10	0.2243	0.262	42	0.487	0.792
11	0.3238	0.0449	43	0.4587	0.503
12	0.3298	0.179	44	0.4832	0.524
13	0.3315	0.2[a]	45	0.4334	0.54
14	0.2973	0.3047	46	0.445	0.468
15	0.3387	0.107	47	0.4639	0.848
16	0.3625	0.447	48	0.4651	0.556
17	0.3398	0.412	49	0.461	0.648
18	0.3378	0.313	50	0.4007	0.6817
19	0.3045	0.3037	51	0.4242	0.6[a]
20	0.417	0.7238	52	0.4557	0.418
21	0.4135	1.0598	53	0.4187	0.3606
22	0.4267	1.0935	54	0.38	0.3643
23	0.377	0.5[a]	55	0.5562	0.829
24	0.3508	0.793	56	0.5676	0.829
25	0.3937	0.5[a]	57	0.5155	0.6[a]
26	0.3911	0.603	58	0.5267	0.609
27	0.3521	0.495	59	0.486	0.6[a]
28	0.4031	0.854	60	0.5676	1.0164
29	0.382	0.6339	61	0.5331	1.2227
30	0.3165	0.8202	62	0.4685	1.1478
31	0.4024	0.562	63	0.529	0.728

[a] Estimated relative response factor based on other isomeric PCBs

Table 1 (continued)

Isomer[P] No.	Relative Retention time	Relative response factor	Isomer[P] No.	Relative Retention time	Relative response factor
64	0.4999	0.607	115	0.6171	1.1328
65	0.4671	0.8408	116	0.6132	1.3987
66	0.5447	0.646	117	0.615	0.8895
67	0.5214	0.6[a]	118	0.6693	0.87
68	0.504	0.726	119	0.5968	0.8239
69	0.451	0.8024	120	0.6256	0.7444
70	0.5407	0.658	121	0.5518	0.7659
71	0.4989	0.468	122	0.6871	0.7247
72	0.4984	0.5515	123	0.6658	0.6645
73	0.4554	0.5805	124	0.6584	0.848
74	0.5341	0.671	125	0.6142	0.556
75	0.4643	0.6461	126	0.7512	0.4757
76	0.5408	0.5795	127	0.7078	0.5834
77	0.6295	0.3812	128	0.7761	1.188
78	0.6024	1.1151	129	0.7501	0.997
79	0.5894	0.881	130	0.7284	0.952
80	0.5464	0.7278	131	0.6853	0.8492
81	0.6149	0.7159	132	0.7035	0.7303
82	0.6453	0.773	133	0.6871	1.148
83	0.6029	0.6339	134	0.6796	0.7331
84	0.5744	0.386	135	0.6563	0.7031
85	0.6224	0.7396	136	0.6257	0.444
86	0.6105	0.7968	137	0.7329	1.112
87	0.6175	1.021	138	0.7403	0.827
88	0.5486	0.6892	139	0.6707	0.7219
89	0.5779	0.561	140	0.6707	0.6732
90	0.5814	0.611	141	0.7203	1.352
91	0.5549	0.571	142	0.6848	1.218
92	0.5742	0.5375	143	0.6789	0.7088
93	0.5437	0.6676	144	0.6563	0.8764
94	0.5331	0.4514	145	0.6149	0.6789
95	0.5464	0.443	146	0.6955	0.728
96	0.5057	0.4308	147	0.6608	0.6[a]
97	0.61	0.631	148	0.6243	0.554
98	0.5415	0.6246	149	0.6672	0.572
99	0.588	0.613	150	0.5969	0.5676
100	0.5212	0.5871	151	0.6499	0.785
101	0.5816	0.668	152	0.6062	0.5235
102	0.5431	0.4561	153	0.7036	0.688
103	0.5142	0.6068	154	0.6349	0.57
104	0.4757	0.4561	155	0.5666	0.586
105	0.7049	0.94	156	0.8105	1.389
106	0.668	1.0046	157	0.8184	1.1965
107	0.6628	0.8183	158	0.7429	1.132
108	0.6626	2.0654	159	0.7655	0.9934
109	0.6016	0.9625	160	0.7396	1.1914
110	0.6314	0.65[a]	161	0.6968	0.9672
111	0.6183	0.6601	162	0.7737	1.0322
112	0.5986	0.8286	163	0.7396	0.9976
113	0.5862	0.604	164	0.7399	0.9948
114	0.6828	1.0261	165	0.692	1.0777

Table 1 (continued)

Isomer[P] No.	Relative Retention time	Relative response factor	Isomer[P] No.	Relative Retention time	Relative response factor
166	0.7572	1.0421	188	0.692	0.7337
167	0.7814	1.0658	189	0.9142	1.5091
168	0.7068	0.8375	190	0.874	1.31
169	0.8625	0.8355	191	0.8447	1.4741
170	0.874	0.75	192	0.8269	1.599
171	0.8089	1.1712	193	0.8397	1.4167
172	0.8278	1.172	194	0.962	1.868
173	0.8152	2.044	195	0.9321	0.415
174	0.7965	0.806	196	0.8938	1.2321
175	0.7611	0.381	197	0.8293	0.9522
176	0.7305	1.0589	198	0.8845	1.07
177	0.8031	1.009	199	0.8494	1.1508
178	0.7537	0.621	200	0.8197	0.369
179	0.7205	0.8237	201	0.8875	0.803
180	0.8362	1.295	202	0.8089	1.165
181	0.7968	1.6046	203	0.8938	1.629
182	0.7653	1.1272	204	0.8217	0.8034
183	0.772	0.976	205	0.9678	1.406
184	0.7016	1.0046	206	1.0103	1.673
185	0.7848	1.437	207	0.9423	1.3257
186	0.7416	1.2236	208	0.932	1.1756
187	0.7654	1.122	209	1.0496	1.139

4 Applications of High Resolution Gas Chromatographic Analysis

Table 2 summarizes the quantitative and qualitative analysis of PCBs in Aroclor 1260 and a composite human milk sample from the State of Michigan (the chromatograms are illustrated in Figures 2 and 5) (42). These results represent the first report of the congener-specific analysis of PCBs in a commercial and environmental sample.

Several PCB congeners identified in the milk sample, including 2,2′,4,4′,5,5′-hexa- (no. 153), 2,2′,3,4,4′,5′-hexa- (no. 138), 2,2′,3,3′,4,4′,5-hepta- (no. 170), and 2,2′,3,4,4′,5,5′-hepta- (no. 180), are major components of Aroclor 1260. 2,3,3′,4,4′,5-Hexachlorobiphenyl (no. 156) is also a major PCB component of the human milk extract but is a minor component of Aroclor 1260. These five PCB congeners possess several common structural features including (i) six or more chlorine atoms per biphenyl moiety and (ii) the presence of only four different substitution patterns (i.e., 3,4-, 2,4,5-, 2,3,4,5-, and 2,3,4-) on both phenyl rings. PCBs no. 153 and 180 do not contain adjacent unsubstituted carbon atoms and are therefore resistant to metabolic breakdown (53), and their persistence in human tissues was not unexpected. The results also suggest that the higher chlorinated PCBs (no. 138 and 170) that contain a 2,3,4-trichlorophenyl group are also

Table 2. Quantitative and qualitative analysis of PCBs in Aroclor 1260 and a human breast milk extract

Congener Name[a]	Percentage in Aroclor 1260	Percentage in human milk[b]	Congener Name[a]	Percentage in Aroclor 1260	Percentage in human milk[b]
PCB-018	0.12	–	PCB-118	0.49	6.5
PCB-017	0.05	–	PCB-134	0.35	–
PCB-024	0.01	–	PCB-114	–	0.33
PCB-016	0.04	–	PCB-131	0.07	–
PCB-029	0.02	–	PCB-122	0.12	0.53
PCB-026	0.02	–	PCB-146	1.3	1.9
PCB-028	0.04	8.8	PCB-153	9.6	12.0
PCB-021	0.01	–	PCB-141	2.5	0.29
PCB-033	0.09	2.2	PCB-176	0.33	–
PCB-053	0.04	–	PCB-137	0.22	0.87
PCB-022	0.01	0.65	PCB-130	–	0.59
PCB-045	0.07	–	PCB-138	6.5	10.0
PCB-046	0.02	0.25	PCB-158	0.70	0.55
PCB-052	0.25	1.9	PCB-129	0.20	–
PCB-043	0.02	–	PCB-178	1.2	–
PCB-049	0.06	0.66	PCB-175	0.49	–
PCB-048	0.29	0.37	PCB-187	4.5	1.5
PCB-044	0.11	0.78	PCB-183	2.3	1.4
PCB-037	0.04	2.9	PCB-128	0.47	0.33
PCB-042	0.04	–	PCB-167	0.16	0.85
PCB-041	0.25	1.3	PCB-185	4.1	0.11
PCB-040	0.03	–	PCB-174	5.5	0.39
PCB-100	0.02	–	PCB-177	1.9	0.61
PCB-074	0.03	11.0	PCB-171+202	1.2	0.37
PCB-070+076	0.15	0.61	PCB-156	0.45	4.87
PCB-095	2.7	–	PCB-173	0.06	–
PCB-091	0.07	–	PCB-200	0.78	–
PCB-056+060	0.14	0.71	PCB-157	–	0.47
PCB-084	0.65	–	PCB-172	0.78	0.31
PCB-101	2.5	0.97	PCB-180	9.1	5.3
PCB-099	0.13	4.8	PCB-193	0.47	0.19
PCB-119	–	0.08	PCB-191	0.10	0.90
PCB-083	0.04	–	PCB-199	0.33	–
PCB-097	0.45	–	PCB-170	6.8	5.3
PCB-087	0.45	0.82	PCB-201	2.9	0.85
PCB-085	0.13	–	PCB-203	3.1	0.79
PCB-136	1.4	–	PCB-196	2.5	0.18
PCB-110	1.7	1.0	PCB-189	0.15	2.4
PCB-154	0.02	–	PCB-195	3.1	0.31
PCB-082	0.11	–	PCB-207	0.08	–
PCB-151	2.5	0.59	PCB-194	1.7	0.48
PCB-144+135	1.5	0.51	PCB-205	0.11	0.06
PCB-107	0.03	0.31	PCB-206	0.85	0.24
PCB-149	7.4	–	PCB-209	0.06	0.09

[a] Congener names adapted from Ballschmiter, K., and Zell, M., Fresenius, Z. Anal. Chem., Vol. 302, 20–31 (1980)
[b] Human milk sample collected and extracted by Michigan Department of Public Health under Cooperative Agreement CR807192 with the Large Lakes Research Station, U.S. Environmental Protection Agency

resistant to metabolism and environmental breakdown and readily bioaccumulate in human tissues. The persistence of 2,3,3',4,4',5-hexachlorobiphenyl in human tissues has previously been reported (54–56) and is somewhat surprising
since 3,4-dichloro substituted PCBs are readily metabolized (53).

The four remaining PCB congeners identified in the human milk extract, milk,
namely 2,4,4'-tri (#28), 2,4,4',5-tetra- (#24), 2,2',4,4',5-penta (#99) and 2,3',4,4',5-
penta (#118) are minor components of Aroclor 1260. It is likely that these penta-
trichlorinated biphenyl congeners are derived from the lower chlorinated PCB
formulations; however, it is noteworthy that with the exception of 2,4,4'-trichlo-
robiphenyl, all of these compounds contain a 2,4,5-trichloro substitution pattern
on one of the phenyl rings and a p-chloro group on the second phenyl ring. This
high-resolution analytical study also identified 2,4,4'-trichlorobiphenyl as a major PCB component and confirmed a previous report that this compound was
present in a Japanese human milk extract (55). The reasons for the persistence of
the lower chlorinated PCBs, particularly PCB #28, in human tissues are unknown. It was also of interest to note that several PCB congeners which are major
components (22.8% of total) of Aroclor 1260 [2,2',3,5',6-penta- (2.7%);
2,2'3,4',5',6-hexa- (7.4%); 2,2',3,4,5,5',6-hepta- (4.1%); 2,2',3,3',4,5,6'-hepta-
(5.5%); 2,2'3,3',4,4',5,6-octachlorobiphenyl (3.1%)] are present in only trace
levels (0.81%) in the human milk extracts.

With the exception of 2,2'3,3',4,4',5,6-octachlorobiphenyl, all of these compounds possess a 2,3,6-trichloro- or 2,5-dichloro-substition pattern on at least
one of their phenyl rings, and because of the two adjacent unsubstituted carbon
atoms rapid metabolic degradation of these congeners would be expected (53).
The failure of the octachlorobiphenyl to accumulate in the human may be due to
uptake factors since the highly persistent higher chlorinated PCBs ($Cl_8–Cl_{10}$) are
not major contaminants in human tissues.

5 High Resolution PCB Analysis: Implications for Risk Assessment

Subsequent chapters in this book describe studies which demonstrate that the biologic and toxic effects of PCBs are highly structure-dependent (57–59). For example, the most toxic group of PCBs, namely 3,4,4',5-tetra-, 3,3',4,4'-tetra-,
3,3',4,4',5-penta- and 3,3',4,4',5,5'-hexachlorobiphenyl are coplanar and are approximate isostereomers of 2,3,7,8-tetrachlorodibenzo-p-dioxin (TCDD). The
eight monoortho analogs of the coplanar PCBs namely 2,3,4,4',5-, 2',3,4,4',5-,
2,3,3',4,4'- and 2,3',4,4',5-penta-, 2,3,3',4,4',5-, 2',3,4,4',5,5'- and 2',3,3',4,4',5'-
hexa-, and 2,3,3',4,4',5,5'-heptachlorobiphenyl, are also moderately toxic and
elicit the same spectrum of effects reported for the more toxic coplanar PCBs.
Based on structure-activity considerations, the remaining 197 PCB congeners are
expected to be relatively non-toxic with respect to their "TCDD-like" activities.
Therefore, environmental risk assessment for PCBs must not only take into account overall "total PCB" residue levels but also the concentrations of individual
toxic PCBs present in these mixtures. This can only be accomplished by high res-

olution capillary gas chromatography with the appropriate reference standards. Since the uptake and persistence of individual PCBs is highly matrix dependent, the potential "risk" from PCBs will be dependent on the specific matrix in question and the levels of the toxic congeners present in these matrices.

Current PCB regulations for fish, wildlife and food products are determined by "estimated total PCB levels" whereas an alternative approach, based on the levels of toxic PCB congeners present in a regulated product can now be developed (60, 61). This approach is presently feasible since the toxic PCB congeners have been identified (57–59) and these compounds can be unambiguously quantitated in environmental samples using the high resolution analytical techniques described in this chapter.

6 Acknowledgements

The financial assistance of the Texas Agricultural Experiment Station, the Environmental Protection Agency and the Chester Reed Endowment are gratefully acknowledged.

7 References

1. Hutzinger O, Safe S, Zitko V The Chemistry of PCBs. CRC Press, Cleveland, Ohio
2. Brinkmann UA Th, De Kok A In: Kimbrough RD (ed) (1980) Halogenated Biphenyls, Terphenyls, Naphthalenes, Dibenzodioxins and Related Products. Elsevier/North Holland, Amsterdam, 1
3. Thomann RV, Mueller JA In: Mackay D, Paterson S, Eisenreich SJ, Simmons MS (eds) (1983) Physical Behavior of PCBs in the Great Lakes. Ann Arbor Science, Ann Arbor, 283
4. Eisenreich SJ, Looney BB In: Mackay D, Paterson S, Eisenreich SJ, Simmons MS (eds) (1983) Physical Behavior of PCBs in the Great Lakes. Ann Arbor Science, Ann Arbor, 141
5. Krauss PB, Suns K, Johnson AF In: Mackay D, Paterson S, Eisenreich SJ, Simmons MS (eds) (1983) Physical Behavior of PCBs in the Great Lakes. Ann Arbor Science, Ann Arbor, 385
6. Sugiura K, Kitamura M, Matsumoto E, Goto M (1986) Arch. Environ. Cont. Toxicol. *15*:69
7. Tanabe S, Hidaka H, Tatsukawa R (1983) Chemosphere *12*:277
8. Kutz FW, Strassman SC, Sperling JF (1979) Ann. N.Y. Acad. Sci. *320*:60
9. Wasserman M, Wasserman D, Cucos S, Miller H (1979) J. Ann. N.Y. Acad. Sci. *320*:69
10. Stratton CL, Sosebee JL (1976) Environ. Sci. Technol. *10*:1229
11. Buckley EH In: (1982) Science. Washington, D.C. *216*:520
12. Tanabe S, Hidaka H, Tatsukawa R (1983) Chemosphere. *12*:277
13. Risebrough RW, Reich P, Peakall DB, Herman SG, Kirven MN In: (1968) Nature. London, *220*:1098
14. Atlas E, Giam CS (1981) Science. Washington, D.C., *211*:163
15. Peakall DB (1975) CRC Crit. Rev. Environ. Control. *5*:469
16. Nisbet ICT, Sarofim AF (1972) EHP, Environ. Health Perspect. *1*:21
17. Landrigan PJ In: Kimbrough RD (ed) (1980) Halogenated Biphenyls, Terphenyls, Naphthalenes, Dibenzodioxins and Related Products. Elsevier/North Holland, Amsterdam

18. Mes J, Campbell DS, Robinson RN, Davies DJA (1977) Bull. Environ. Contam. Toxicol. *17*:196
19. Jensen S, Sundstrom G (1974) Ambio. *3*:70
20. Watanabe L, Yakushiji T, Kuwabara K, Yoshida S, Maeda K, Kashimoto T, Koyama K, Kunita N (1979) Arch. Environ. Contam. Toxicol. *8*:67
21. Barbehenn KR, Reichel WL (1981) J. Toxicol. Environ. Health *8*:325
22. Bjerk JE, Brevik EM (1980) Arch. Environ. Contam. Toxicol. *9*:743
23. Holdrinet MVH, Braun HE, Frank R, Stopps GJ, Smout MS, McWade JW (1977) Can. J. Public Health *68*:74
24. Safe S (1982) Toxicol. Environ. Chem. *5*:153
25. Niemi GJ, Davis TE, Veith GD, Vieux B (1986) Arch. Environ. Contam. Toxicol. *15*:313
26. Bush B, Simpson KW, Shane L, Koblintz RR (1985) Bull. Environ. Contam. Toxicol. *34*:96
27. Bush B, Snow J, Connor S, Rueckart L, Han Y, Dymerski P, Hilker D (1983) Arch. Environ. Contam. Toxicol. *12*:739
28. Ballschmiter K, Buchert H, Bihler S (1981) Fres. Zeit. Anal. Chem. *306*:323
29. Brunn H, Manz D (1982) Bull Environ. Contam. Toxicol. *28*:599
30. Zabik ME, Merrill C, Zabik MJ (1982) Bull Environ. Contam. Toxicol. *28*:710
31. Bush B, Connor S, Snow J (1982) J. Assoc. Offic. Anal. Chem. *65*:555
32. Schulte E, Malisch R (1983) Fres. Zeit. Anal. Chem. *314*:545
33. Cooper SD, Moseley MA, Pellizzari ED (1985) Anal. Chem. *57*:2469
34. Bush B, Murphy MJ, Connor S, Snow J, Barnard E (1985) J. Chrom. Sci. *23*:509
35. Mullin M, Sawka G, Safe L, McCrindle S, Safe S (1981) J. Anal. Toxicol. *5*:138
36. Mullin MD, Pochini CM, McCrindle S, Romkes M, Safe SH, Safe LM (1985) Environ. Sci. Technol. *18*:468
37. Kerkhoff MAT, de Vries A, Wegman RCC, Hofstee AMW (1982) Chemosphere. *11*:105
38. Albro PW, Corbett JT, Schroeder JL (1981) J. Chrom. *205*:103
39. Sissons D, Welti D (1971) J. Chrom. *60*:15
40. Strang C, Levine SP, Orlan B, Gouda TA, Saner WA (1984) J. Chrom. *314*:482
41. Wolff MS, Thornton J, Fischbein A, Lilis R, Selikoff IJ (1982) Toxicol. Appl. Pharmacol. *62*:294
42. Safe S, Safe L, Mullin M (1985) J. Agric. Food Chem. *33*:24
43. Sawyer L (1978) J. Assoc. Offic. Anal. Chem. *61*:282
44. Webb RG, McCall AC (1973) J. Chrom. Sci. *11*:366
45. Pellizzari ED, Moseley MA, Cooper SD (1985) J. Chrom. *334*:277
46. Newton DA, Laski R (1983) J. Chrom. Sci. *21*:161
47. Wolff MS, Fischbein A, Rosenman KD, Levin SM (1986) Chemosphere *15*:301
48. Slivon LE, Gebhart JE, Hayes RL, Alford-Stevens AL, Budde WL (1985) Anal. Chem. *57*:2464
49. Alford-Stevens AL, Bellar TA, Eichelberger JW, Budde WL (1986) Anal. Chem. *58*:2014
50. Hansen LG (1979) Ann. N.Y. Acad. Sci. *320*:183
51. Harvey RG, Steinhauer WG (1974) Atmos. Environ. *8*:777
52. Ballschmiter K, Zell M (1980) Fres. Zeit. Anal. Chem. *302*:20
53. Matthews HB, Dedrick RL (1984) Annu. Rev. Pharmacol. Toxicol. *24*:85
54. Yoshihara S, Kawano K, Yoshimura H, Kuroki H, Masuda Y (1979) Chemosphere *8*:531
55. Yakushiji T, Watanabe I, Kuwabara K, Yoshida S, Koyama K, Kumita N (1979) Int. Arch. Occup. Environ. Health *43*:1
56. Fujiwara K (1975) Sci. Total Environ. *4*:219
57. Safe S (1984) CRC Crit. Rev. Toxicol. *13*:319
58. Safe S, Bandiera S, Sawyer T, Robertson L, Safe L, Parkinson A, Thomas PE, Ryan DE, Reik LM, Levin W, Denomme MA, Fujita T (1985) Environ. Health Persp. *60*:47
59. Leece B, Denomme MA, Towner R, Li SMA, Safe S (1985) J. Toxicol. Environ. Health *16*:379
60. Strang C, Levine SP, Orlan BP, Gouda TA, Saner WA (1984) J. Chrom. *314*:482
61. Stalling DL, Hackens JN, Petty JD, Johnson JL, Saunders HO (1979) Ann. N.Y. Acad. Sci. *320*:48

Environmental Toxicology of Polychlorinated Biphenyls

L. G. Hansen[1]

Over 10^9 Kg of PCB have been produced world-wide in the last 5 decades and about one-third of this is estimated to be in mobile environmental reservoirs. Entropic dispersion is decreasing geographic distinctions in residue levels, but hazards still persist in spite of dramatic decreases in production and release.

The complex nature of PCBs and the various forces acting unequally on components of PCB mixtures make it difficult to firmly establish trends in disposition. The entire spectrum of potential PCB effects is probably not known and many demonstrated effects are frequently discounted because they are expressed inconsistently or are seemingly ambiguous. This along with the complexity of the environmental residues makes it difficult to accurately assess the potential hazards.

In order to adequately address the potential as well as real hazards of the global PCB burden, the full complexity of PCB disposition and toxicity must be re-recognized. This article attempts to integrate the more dramatic observations with the more subtle and more common manifestations of PCB exposure.

[1] College of Veterinary Medicine and Institute for Environmental Studies, University of Illinois, Urbana, IL 61801, USA

1 Introduction

Polychlorinated biphenyls (PCBs) have been produced commercially since before 1930. They proved to be highly versatile mixtures and their uses continued to expand during the early 1970's even after the unanticipated world-wide environmental contamination had been discovered (Jensen et al., 1969; Koeman et al., 1969). Over 600,000 metric-tons were produced and/or used in the U.S. during this time and it is estimated that worldwide production totaled about 1,200,000 metric-tons (Table 1).

With low acute toxicities (Fishbein, 1974), these mixtures were considered generally biologically inactive even though industrial exposure had demonstrated hepatic and dermatological effects (Fishbein, 1974; Hansen, 1987). Thus, use and disposal were not carefully monitored and it is estimated that one-third of the world-wide production of PCBs has been released into the global environment (Table 1).

Table 1. Estimated production and disposition of PCBs

	U.S.[a]	Worldwide[b]
Production/use	610×10^6 kg	1200×10^6 kg
Mobil environmental reservoir	82	400
Static reservoirs		
In service	340	–
Dumps	130	–
Total static	470	800

[a] NAS, 1979
[b] Tatsukawa and Tanaba, 1984

2 Environmental Distribution

Many countries now impose strict controls on the use and release of PCBs. Release into the environment has declined dramatically in the last decade, but continued release from reservoirs (Table 1) into burdened ecosystems (Table 2) appears inevitable for several more decades (Barros et al., 1984). While ocean and lake sediments appear to be the ultimate sinks for environmental PCBs (NAS, 1979; Tatsukawa and Tanabe, 1984; Eisenreich and Johnson, 1983), they also serve as sources due to natural phenomena (Swain, 1983; Hallett, 1984); the greater burden of superficial lake sediments (Eisenreich and Johnson, 1983) and coastal sediments (NAS, 1979; Sugiura et al., 1986; Tatsugawa and Tanabe, 1984), presents an opportunity for more intense re-release due to geophysical and anthropogenic disturbances.

Table 2. Summary of estimated PCBs in mobile environmental reservoirs[a]

Reservoirs	Estimated PCB (10^6 kg)		
	U.S.-North Atlantic	World[b]	
Atmosphere	1.8	–	
Lithosphere	2.8	(5.5)[c]	
Freshwater			
Sediment	7.1		
Water	0.035		
Biota	0.030		
Marine		Open ocean	Coastal[c]
Sediment	2.7	0.10	
Water	66	230	
Biota	0.030	0.27	
Sewage sludge	4.8		
Totals	85.3	230	(138)[c]

[a] NAS, 1979. High estimates were used because the total of the low estimates was 10-fold below the NAS estimated total reservoir, whereas, the total of the high estimates was essentially the same
[b] Tatsukawa and Tanaba, 1984
[c] 12% of use (1200×10^6 kg) estimated by authors in terrestrial and coastal. Using NAS (0.46% in Lithosphere) leaves 11.5% for coastal residues; a higher percentage would be in coastal sediments and biota due to lower water volume and proximity to sources. NAS (1979) also infers most sediment PCB is coastal

Moderately contaminated lake sediments can be dredged without release of PCB into the food web, although some mixing into lower levels is observed (Sondergren, 1984). A highly localized spill of Aroclor 1242 was effectively (90%) recovered from the Duwamish River in Puget Sound, also with more positive than negative results (Pavlou and Hom, 1979). The dredging of large areas of highly contaminated sediments, however, must be carefully considered since, even meticulous removal of the sediment may create additional hazards in transport, storage and decontamination. Sediments, containing less than 10 ppm PCB can be considered less of an environmental hazard when applied to distressed lands than when left in aquatic systems such as navigation channels (Van Luik, 1984).

Atmospheric burdens and fluxes insure that no geographical area will be spared some level of PCB contamination and this may account for the redistribution of PCBs, e.g., from the lower Great Lakes to the upper nonindustrialized Great Lakes (Hallett, 1984). The areas of most concern include the open ocean, which contains the greatest amount of mobile PCB; aquatic systems in which residues are less stable and more subject to mobilization; dumps and equipment in service, which could result in large accidental releases; and isolated terrestrial hot spots where hazard may be more limited in scope but more intense.

The intense areas of terrestrial exposure are seldom addressed in discussions of global PCB patterns because of the relatively small contribution to total reservoirs and, perhaps, a feeling of more secure containment. In 1977, concentrations of PCB in or on vegetation near PCB-containing landfills in New York (representing less than 0.2% of the PCB in this category for the US) averaged from 100–500 ppm (dry weight) with a level as high as 2770 ppm recorded (Horn et al., 1979). Contamination was primarily due to surface adsorption of dust and deposition from high atmospheric levels. Control concentrations (vegetation collected several miles west) were about 0.2 ppm, while vegetation grown on Hudson River dredge spoil disposal sites contained about 0.6 ppm. By 1979, 2 years after measures to control PCB emissions were instituted, levels in upstate New York, hay were about 0.08–0.10 ppm (dry weight), with other declines apparent between 1978 and 1980 (Eisler, 1986). More recent studies indicate PCB concentrations of 0.01–0.2 ppm (fresh weight) in leaves from purple loosestrife (*Lythrum salicaria*) growing in a contaminated soil (ca. 0.12 ppm) and that the major route of entry was absorption via roots (Bush et al., 1986). It is difficult to estimate the effects of chronic consumption of the amounts of PCB associated with these plant materials, however the contamination is declining. Herbivores in undisturbed habitats (desert Bighorn sheep, *Ovis canadensis*) still have chlorinated hydrocarbon residues of 0.3–1.0 ppm on a fat basis; the majority of the residue is PCB (Turner, 1978).

Sewage sludge represents another large reservoir of PCB (Table 2), that frequently finds its way into the terrestrial environment (Hansen et al., 1981; Hansen and Chaney, 1984). The concentrations of PCB found, however, do not seem to pose an immediate hazard (Hansen and Chaney, 1984; Hansen, 1987) even to home gardeners using PCB-enriched sludge as a soil amendment (Baker et al., 1980). Other toxicants present, such as Cd, are a greater health hazard but the possibility of toxic interactions is possible. In addition, the presence of this reservoir and potential for dispersion will slow the decline in ambient global PCB levels.

An even more limited and more intense source of lithospheric hot spots are PCB-coated silos, however the problem is being addressed. From the early 1940's through about 1970, certain sealants for concrete silos contained large amounts of Aroclor 1254 (U.S.) or Aroclor 1260 (Europe) (Skrentny et al., 1971; Alencastro et al., 1984). PCBs were leached from the walls by organic acids during ensilage, transferred to cattle in the silage and into the human food chain by meat and milk. PCB concentrations in soil near the silos over 30,000 ppm have been reported as well as over 1 ppm in house dust (Hansen, 1987). Significant increases in cancer incidence (Humphrey, 1983) and other toxicoses as well as serum PCB levels higher than accidental poisoning victims (Hansen, 1987) have been observed in small groups of families associated with silo farms. Most of these silos have been closed since the mid-1970's and many are being dismantled and buried. Monsanto Chemical Company was reportedly not aware of this application of their product for 25 years and documentation regarding this use of PCB is incomplete (Pappageorge, 1983). Thus, the total amount of PCB used for silo coatings is not available but it cannot be large in proportion to total PCB use and manufacture; however, the extent of dispersion via milk and the human cost are disproportionately large.

More detailed and thoroughly discussed accounts of global PCB disposition in the atmosphere and hydrosphere can be found in the references cited previously.

The estimates given in Tables 1 and 2 are guides based on sound assumptions. Specific numbers would be useful for developing concepts as well as predicting and/or confirming special problems and trends; nevertheless, the erratic evolution and application of sampling and analytical techniques makes it difficult to establish unequivocal comparisons.

Certain generalizations have been extracted from the previous reviews which in turn have examined extensive data and addressed the ambiguities. The major thrust of this review is to focus on the effects of PCBs on the biota, so that the following generalizations will be useful in establishing perspectives upon which to develop concepts of the environmental toxicology of PCBs.

1. The atmosphere, although containing a relatively low percentage of the mobile environmental PCB reservoir, is the most dynamic reservoir and accounts for much movement among reservoirs.
2. Average atmospheric PCB concentrations in metropolitan areas (ca. 5 ng/m^3) are much higher than average concentrations in rural and oceanic atmospheres (ca. 0.05 ng/m^3). Average levels of 1000 ng/m^3 have been observed near manufacturing plants.
3. Estimation of PCB in the lithosphere is very uncertain, but localized areas of high contamination are of great concern. (Note previous discussion and that point 4 (below) refers to total global PCB).
4. Current landfills and equipment dumps do not appear to be contributing greatly to global mobile environmental PCB. Some volatilization from landfills occurs, but little groundwater contamination has been demonstrated (NAS, 1979).
5. In areas of low contamination, volatilization from organic soils is negligible and volatilization from non-organic surfaces appears to be in equilibrium with atmospheric fallout.
6. Lithospheric biota including wildlife, livestock, humans and, especially (due to biomass), plants contain more PCB than the medium (top cm of soil).
7. Land runoff, direct discharge and, especially, atmospheric deposition contribute most to PCB loads in the hydrosphere.
8. Accounting for over 95% of the surface freshwater in the U.S., the Great Lakes contribute over 97% of the U.S. aquatic PCB load. The St. Lawrence drainage is a major force in moving part of this burden into the North Atlantic.
9. With reduced discharges, atmospheric deposition currently accounts for 60–80% of the movement into the Great Lakes as well as transfer to oceans and redistribution from the lower lakes to the upper, less industrialized lakes.
10. Other aquatic reservoirs such as the Hudson River create serious local problems even though the entire Appalacian-Atlantic coast drainage contributes less than 1% of the U.S. hydrospheric burden.
11. The water of the open ocean, although low in PCB concentration relative to some lakes and most coastal areas, contains the greatest total amount of the mobil environmental PCB reservoir. Concentrations are highest in the mid-latitudes of the northern hemisphere, vaguely correlating with industrial development.

12. In spite of lower total amounts in freshwaters, freshwater sediments contain more PCB and freshwater biota contain about the same amount of PCB as the respective larger oceanic compartments.
13. Ocean residues are in flux with atmospheric and sediment PCBs.
14. Sediments in oceans and large lakes are the ultimate sinks for mobile PCBs, although re-introduction into other compartments is possible. PCBs in sediments of rivers, estuaries and some coastal areas act more as PCB sources than sinks.
15. Sedimentation rates are more rapid in higher latitude (eutrophic)ocean waters due to greater particulate matter. Higher chlorinated congeners bind more readily to particulates and, thus, sediment more rapidly.
16. PCBs are complex mixtures of some 30–60 congeners which are the major PCB components of most environmental extracts and 60–90 less common congeners. Each individual compound exhibits a unique combination of physical, chemical and biological properties.

3 Composition of Residues

The last general point requires considerable amplification in order to maintain useful perspectives on the environmental toxicology of PCBs. Fortunately, several aspects are addressed in other chapters of this volume.

There are 209 possible structures for the biphenyl molecule containing from 1 to 10 chlorine atoms. Commercial production involved random chlorination, but principles of chemical substitution favored certain structural patterns. Although commercial PCB mixtures consist of congeners dissolved in each other that tend to move in concert, they actually move individually in the environment and each congener has properties adequately distinct to favor unequal disposition. With the multitude of physical, chemical and biological forces acting in the environment, it is not surprising that various residues differ slightly to dramatically in congener composition.

The variable composition of PCB residues has three main consequences. Firstly, analytical methodologies and quantitation procedures can be quite complicated; attempts to streamline these techniques led to inconsistent and frequently incomparable results from laboratory to laboratory and from time to time as techniques were modified. Secondly, different mixtures of congeners tend to mobilize differently, partition into different phases, and have different degrees of persistence/degradability. Thirdly, the biological activity differs markedly, both qualitatively and quantitatively among isomers as well as congeners with different degrees of chlorination.

3.1 Analytical Consequences

It is not possible to adequately address the problems of sampling and analysis, and the implications of variations in the evolution of techniques. The importance

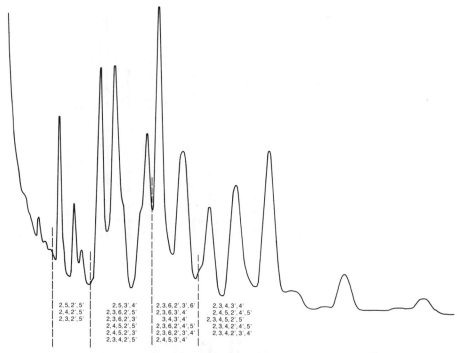

2,5,2',5'	2,5,3',4'	2,3,6,2',3',6'	2,3,4,3',4'
2,4,2',5'	2,3,6,2',5'	2,3,6,3',4'	2,4,5,2',4',5'
2,3,2',5'	2,3,6,2',3'	3,4,3',4'	2,3,4,5,2',5'
	2,4,5,2',5'	2,3,6,2',4',5'	2,3,4,2',4',5'
	2,4,5,2',3'	2,3,6,2',3',4'	2,3,4,2',3',4'
	2,3,4,2',5'	2,4,5,3',4'	

Fig. 1. Representative packed column glc tracing of Aroclor 1254. Congeners which would elute with similar retention times are given below the peaks

of defining residue composition was recognized soon after PCB ubiquity was discovered (e.g., Hutzinger et al., 1974; Webb and McCall, 1972; Jensen and Sundström, 1974). Nevertheless, the majority of environmental residues have been reported as total PCB derived from estimation procedures based on characteristic packed column gas chromatographic peaks (Figure 1). The complexities of data reporting and interpretation, as much as analytical deficiencies, made these total PCB estimations almost essential for world PCB inventories and trends; however, it must be recognized that such data can neither be used literally nor compared directly without proper consideration of the techniques employed (e.g., Mes et al., 1980; Burse et al. 1983; Tuinstra, 1984; Klein, 1984; Jensen, 1984; Bedard et al., 1986). Analyses quantifying all significant peaks for total PCB determination are more accurate reflections of the total amount and provide more information regarding congener composition (e.g., Bush et al., 1974; Jensen and Sundström, 1974; Collier et al., 1976; Hansen et al., 1976).

Figures 2 and 3 illustrate useful comparisons between Aroclors and among species in relative residue composition by packed-column glc (retention times relative to p,p'-DDE = 100). Early eluting (lower chlorinated) components are generally less persistant, except peaks 37 and 48 (PCBs 28 and 52, respectively); thus, the profile of Aroclor 1242 is altered more dramatically than that of Aroclor 1254. Catfish alter the profile of even Aroclor 1242 very little, even after extended periods of elimination (Hansen et al., 1976).

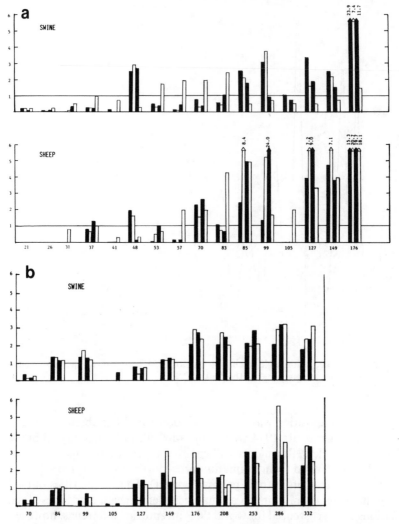

Fig. 2 a, b. Selective accumulation or depletion of PCB peaks in swine and sheep relative to the composition of the parent Aroclor mixture. In the Aroclor administered, all peaks would have a relative value of 1.0. Note that numbers represent relative retention time, not IUPAC congener numbers
a Selective accumulation of Aroclor 1242 components in swine and sheep tissues. Sequence: fat, liver, kidney, blood. **b** Selective accumulation of Aroclor 1254 components in swine and sheep tissues. Sequence: fat, liver, kidney, blood

Fig. 3 a–d. Selective accumulation or depletion of PCB peaks in broiler cockerels and aquatic organisms relative to the composition of the parent Aroclor mixture
a Selective accumulation of Aroclor 1254 components in broiler cockerels. Sequence: fat, liver, kidney, blood, brain. **b** Selective accumulation or Aroclor 1254 components in aquatic organisms. Sequence: algae, mosquito larvae, snail, fish. **c** Selective accumulation of Aroclor 1242 components in broiler cockerels. Sequence: fat, liver, kidney, blood, brain. **d** Selective accumulation of Aroclor 1242 components in channel catfish. Sequence: fat, liver, kidney

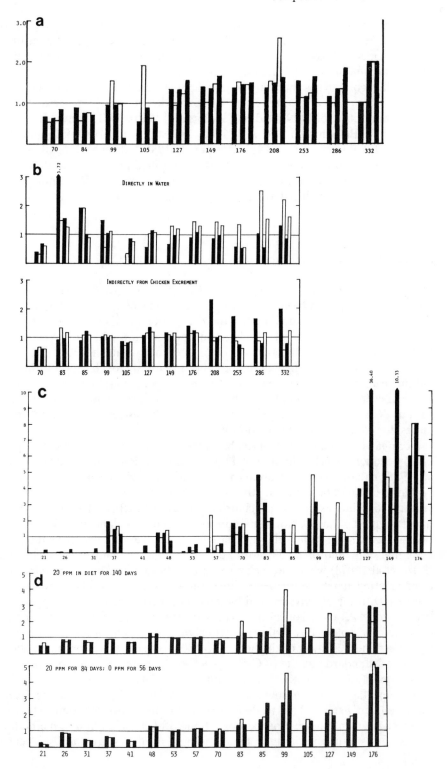

In algae (*Oedegonium*) the profile is changed considerably depending on the source of Aroclor 1254 (Sundlof, 1975); in addition, nutrient enrichment of the aquarium model ecosystem with chicken excrement may contribute to the change in algal accumulation of PCB congeners.

Capillary column gas chromatography (See Chapter 1) permits better congener resolution and, with appropriate standards, can thoroughly characterize the congener composition of PCB residues (Sissons and Welti, 1971; Jensen and Sundström, 1974; Ballschmiter and Zell, 1980; Tuinstra et al., 1980; Zell and Ballschmiter, 1980; Hansen et al. 1981; Bush et al., 1985; Safe et al. 1985 b). This increasing use of specific congener analysis has resulted in more accurate and detailed information permitting more accurate evaluations of PCB problems and trends. It has also produced an immense data resource which has not been thoroughly assessed.

3.2 Disposition and Persistence

Disposition is dependent on physical and chemical properties interacting with physical, chemical and biotic forces. Most pure congeners are solids at temperatures below 40 °C, but some mono- and dichlorobiphenyls have melting points below 30 °C. Their chemical stability and liphilicity are well known properties that contribute greatly to environmental and food chain hazards. PCBs do undergo slow oxidation, hydrolysis and photochemical reductive dechlorination, but the relevance to environmental conditions is uncertain (Hutzinger et al. 1974; Pomerantz et al. 1978; Safe, 1984). Aerobic microbes are capable of metabolizing various PCB congeners (Carey and Harvey, 1978; Bedard et al., 1986), but effects on residue composition may be too subtle to detect.

Acclimated populations of anaerobic bacteria have been reported to be capable of slow reductive dechlorination of PCBs in Hudson River sediments (Brown et al., 1985); enhanced concentrations of lower chlorinated congeners have been reported in comparable Hudson River water, but these may also be attributed to selective solubility rather than dechlorination products (Bush et al., 1985). Potential dissolution of lower chlorinated congeners from the sediments makes their enrichment even more remarkable and the concept of anaerobic dechlorination even more plausible. Photolytic dechlorination preferentially removes *ortho* chlorines (Ruzo et al., 1974) while the acclimated anaerobes remove *meta* and *para* chlorines, creating an *ortho*-enriched residue (Brown et al., 1985). The importance of *ortho* substitution on physical and biological properties will become apparent in later sections. As a word of caution, a cursory examination of the chromatograms presented for Tokyo Bay sediments also show enrichment of lower chlorinated congeners and depletion of the hepta-CB 180; however, this is primarily due to profiling a standard mixture (KC-500 + KC-600) not representative of the likely major input (KC-300 + KC-500) (Sugiura et al., 1986).

As important as water solubility, vapor pressure, and partition coefficients (water/oil and water/air) are to PCB disposition, there does not seem to be good general agreement on these properties. Thoughtful consideration of methodologies as well as results account for, but do not necessarily resolve, the discrepencies in the literature (Tulp and Hutzinger, 1978; Burkhard et al., 1985 a). Reported partition coefficients can differ by 2 orders of magnitude and may differ from calculated log P's by 3 orders of magnitude for the same PCB (Tulp and Hutzinger, 1978). Table 3 presents some ranges of data relevant to solubilities and vapor pressures for various PCBs. Relative GLC retention times should generally agree inversely with relative vapor pressures.

Table 3. Structure and properties of common or commonly studied PCB congeners

PCB no.[a]	Substitution		GLC RRT[b]	Vapor pressure[d] (PAT at 25°C)		Water solubility[e]	
	Ring 1	Ring 2		Exper.	Calculated	Not treated	By elution
4	2	2'	0.2245	175	–	1440	790
15	4	4'	0.3387	2.63	8.99	70	56
18	2,5	2'	0.3378	–	8.97	–	640
28	2,4	4'	0.4031	–	–	85	260
37	3,4	4'	0.4858	–	–	–	–
44	2,3	2'5'	0.4832	–	–	170	–
49	2,4	2'5'	0.461	–	1.13	–	–
52	2,5	2'5'	0.4557	4.97	4.27	46	–
70	2,5	3'4'	0.5407	–	7.69	41	–
77	3,4	3'4'	0.6295[c]	–	–	175	0.75
84	2,3,6	2'3'	0.5744	–	–	–	–
95	2,3,6	2'5'	0.5464	–	–	–	–
97	2,4,5	2'3'	0.61	–	–	–	–
99	2,4,5	2'4'	0.588	–	–	–	–
101	2,4,5	2'5'	0.5816	1.11	–	28	4.2
105	2,3,4	3'4'	0.7049[c]	–	–	–	–
110	2,3,6	3'4'	0.6314	–	–	–	–
118	2,4,5	3'4'	0.6693	–	–	–	–
126	3,4,5	3'4'	0.7512[c]	–	–	–	–
128	2,3,4	2'3'4'	0.7761[c]	–	0.05	–	–
136	2,3,6	2'3'6'	0.6257	–	–	–	–
138	2,3,4	2'4'5'	0.7403	–	–	–	–
141	2,3,4,5	2'5'	0.7203	–	–	–	–
149	2,3,6	2'4'5'	0.6672	–	–	–	–
153	2,4,5	2'4'5'	0.7036	–	0.46	8.8	1.2
156	2,3,4,5	3'4'	0.8105[c]	–	–	–	–
170	2,3,4,5	2'3'4'	0.874[c]	–	–	–	–
180	2,3,4,5	2'4'5'	0.8362	–	–	–	–
183	2,3,4,6	2'4'5'	0.0.772	–	–	–	–
187	2,3,5,6	2'4'5'	0.7654	–	–	–	–
189	2,3,4,5	3'4'5'	0.9142[c]	–	–	–	–
194	2,3,4,5	2'3'4'5	0.962	–	–	7.2	0.27

[a] IUPAC number (Ballschmiter and Zell, 1980)
[b] Retention time relative to octachloronaphthalene (Mullin et al., 1984)
[c] Note effect of limited *ortho* substitution on retention times, e.g., PCB 77 elutes between pentachlorobiphenyls 97 and 110 (as does *p,p'*-DDE)
[d] Burkhard et al., 1985a
[e] Reported by Tulp and Hutzinger (1978) high values may be due to aggregates since water was neither filtered nor centrifuged. Column 2 was elution by water from Florisil

Table 4. Occurrence of various congeners in commercial PCBs and in environmental residues

PCB no.[a]	% Total PCB		Concentration (ng/g)					
	Aroclor 1254[b]	Aroclor 1260[c]	Hudson River[d]		Deep sea fish liver[e]	Fresh-water eel[f]	Eagle liver[g]	Swine fat[h]
			H_2O	Larvae				
4	–	–	0.147	300	–	–	30	BD
15	–	–	0.017	300	–	–	67	BD
18	–	0.12	0.014	1200	–	–	22	–
28	–	0.04	0.008	700	6	35	926	–
44	1.9	0.11	0.002	1200	23	34	1319	BD
49	1.3	0.06	0.008	1800	27	–	–	BD
52	4.2	0.25	0.010	7400	47	110	5230	23
70	3.2	0.15	0.004	2600	18	–	+[g]	–
77	–	–	0.001	60	–	–	–	–
84	(1.7)	0.65	0.001	500	–	–	5712	BD
95	7.6	2.7	–	–	123	130	4245[g]	14
97	2.3	0.45	0.001	600	–	–	4394	BD
99	(2.0)	0.13	0.003	500	–	–	+[g]	–
101	8.8	2.5	0.002	900	200	85	+[g]	7
105	–	–	–	–	129	–	–	–
110	(7.5)	1.7	–	–	–	–	3119	–
118	9.5	0.5	–	–	–	110	3956	10
128	1.5	0.5	0.001	200	44	20	–	18
136	1.5	1.4	–	–	–	–	–	–
138	7.6	6.5	0.002	600	596	200	4838	110
141	1.8	2.5	–	–	15	40	–	BD
149	6.2	7.4	–	–	–	90	–	–
153	6.1	9.6	ND	300	570	180	12869	89
156	0.7	0.45	0.010	100	54	–	–	BD
170	0.7	6.8	–	–	–	30	–	15
180	1.2	9.1	–	–	352	80	–	58
183	–	2.3	–	–	120	–	–	–
187	0.4	4.5	–	–	–	–	–	23
194	–	1.7	–	–	59	–	–	BD
Total	77.4%	62.1%	0.532	66000	2730	–	74090	389

[a] IUPAC number (Ballschmiter and Zell, 1980). See Table 3 for structures. Dashes indicate absence of data rather than absence of congener
[b] Hansen et al., 1983
[c] Safe et al., 1985b
[d] Sedimented water sample and caddis fly *(Hydropsyche leonardi)* larvae from the same location (Bush et al., 1985)
[e] Liver sample from rattail *(Coryphaenoides armatus)* collected from Hudson Canyon, North Atlantic (Stegman et al., 1986)
[f] Tuinstra, 1984
[g] Bald eagle *(Haliaeetus leucocephalus)* from Illinois (Collier et al., 1976). Analysed by packed column glc so that PCB 84 represents 84+99+101; 95 includes 66 and 70; and 97 also includes some 87
[h] Backfat biopsies from sows having foraged intermittently on sewage sludge amended plots (Hansen et al., 1981)
BD = below detection limit of 5 ng/g

The congeners listed are important components of commercial PCBs and PCB residues (Table 4) and further discussion, now, can refer to each by IUPAC number. Indeed, much more information could be gathered for many more congeners, but the data would then be overwhelming. The anticipated inverse relationship between molecular weight and vapor pressure does not occur unless the number of *ortho* chlorines are considered separately (Mullin et al., 1984; Burkhard et al., 1986 b). Note the longer retention times relative to other isomers for PCBs 77, 105, 126, 156 and 189 and disproportionately rapid elution of 136. The planarity of the biphenyl molecule is profoundly influenced by *ortho* substitution; consequently, degree of *ortho* substitution on various isomers influences vapor pressure and glc retention time (Table 3), water solubilities (Tulp and Hutzinger, 1978), retention in fat (Jensen and Sundström, 1974; Kuroki and Masuda, 1977; Sparling and Safe, 1980) and biological activities (Safe, 1984).

Thus, when environmental disposition is discussed, two of the major factors to consider are degree of chlorination and degree of *ortho* chlorination. The third consideration is extent of microbial and macrobiotic metabolism, which is also influenced by the above factors, as well as *para* chlorination (Sipes, 1987). At this time, however, these factors must be considered in relative terms rather than absolute numbers and limited hypotheses must be based on the actual congener composition of environmental residues.

Lower chlorinated biphenyls are more water soluble, even if the progression is not linear due to *ortho* influences. They are leached more readily from silo coatings into silage organic acids, so that aged silos pass progressively less total PCB into the silage (Hansen, 1987). Lower chlorinated biphenyls are also taken up more readily by various terrestrial plants under experimental conditions (Iwata and Gunther, 1976; Suzuki et al., 1977; Eisler, 1986). Most of these studies used very high (e.g., 100 ppm) concentrations in the substrate and lack of detectable residues at 2–3 ppm led Fries (1982) to conclude that most plant contamination results from dust or vaporization of the lower chlorinated congeners.

A lower detection limit permitted characterization of residue composition in contaminated substrate, atmosphere and indigenous plants (Bush et al., 1986). The predominant (more than 5% of residue) lower chlorinated congeners were PCBs 28, 52 and 68 (2,3',4',5) while the very persistent PCBs 128, 138, 153, 170 and 180 ranged from 0.7–2.5% of the plant residue. PCBs 28, 52, 153 and 180 were more significant contributors to the atmospheric residue than 68 (2.7%), 170 (2.3%), 128 (ND) and 138 (ND). These and other relationships indicated systemic absorption as the dominant source of contamination of terrestrial plants, although atmospheric uptake is also significant and becomes dominant in highly contaminated atmospheres. In atmospheres of low PCB content, 2-CB and 2,2'-CB are emitted from contaminated plants (Bush et al., 1986).

Some examples of aquatic, marine and terrestrial residues related to Aroclor composition are given in Table 4. The absence of data does not necessarily mean the congener is absent; it may be below the reliable detection limit or the standard may not have been available. The enrichment of aqueous media with lower chlorinated congeners is a reflection of water solubilities, atmospheric flux, high particle affinities resulting in greater sedimentation rates for higher chlorinated congeners and anaerobic dechlorination in sediments (Brown et al., 1985; Bush et al.,

1985; Tatsukawa and Tanabe, 1984). Biotransformation is facilitated by the presence of two adjacent unsubstituted carbons (vicinyl hydrogens) and absence of *para* substituents (Jensen and Sundström, 1974; Sipes, 1987; Hansen, 1987), so the lower halogenated congeners are generally more rapidly metabolized. Thus, the enrichment of air and water with lower chlorinated congeners is not reflected in higher organisms (Table 4).

The loss of lower chlorinated congeners can also be demonstrated by determining percent chlorine of sequential residues. Atmospheres over the open ocean generally contain a few ppt PCB averaging 40–43% Cl; composition of open ocean waters is somewhat more variable due to atmospheric flux and sedimentation rates, but still more constant than coastal and aquatic systems. In the western North Pacific water, the chlorine content averages 46.3%, indicating significant loss of lower chlorinated compounds or enrichment with those of higher chlorine content. Lower ranking predators, such as squid, accumulate residues of 55–56% chlorine, while striped dolphin residues contain about 58% chlorine (Tatsukawa and Tanabe, 1984).

Livers from the deep-sea fish *Coryphaenoides armatus* (rattail) from an area in the western North Atlantic (Hudson Canyon) proximal to urban centers and dump sites contained 7–8 times more PCB than the same species from a more remote location (Stegeman et al., 1986). The more highly chlorinated congeners clearly predominated (Table 4), and even more so in the rattails from the remote area. Comparing the congener profile with that of caddisfly larvae from the Hudson River indicates that the ocean sediment and the deep ocean food chain contain higher proportions of more highly chlorinated congeners, consistent with the relative sedimentation rates proposed by Tatsukawa and Tanabe (1984). The Hudson River, however, is contaminated primarily with Aroclors 1016, 1221 and 1242. A further consideration is the relative inability of most fish species to metabolize PCB congeners with 4 or more chlorines (Lech and Peterson, 1983).

PCBs 52, 138, 153 and 180 seem to predominate in the residues from most animals (Table 4). PCB 44, the most bioconcentrated congener in caddisfly larvae (Bush et al., 1985) was also present at high levels in the exposed rattail when compared to the rattail from a remote ocean area (38-fold). Accumulation of PCB 44 in eagle liver and swine fat, however, is not remarkable when compared to other tetrachlorobiphenyls (Table 4).

Comparing PCB 44 (2,2',3,5'-tetrachlorobiphenyl) to the prevalent PCB 52 (2,2',5,5'-tetrachlorobiphenyl) indicated a rather constant ratio of 0.38–0.45 for several commercial preparations and this ratio provides a useful marker to evaluate the effects of environmental processes on the relative persistence and uptake of both isomers. PCBs 44 and 52 have similar retention times (vapor pressure), vicinyl hydrogens and intermediate aqueous solubilities. Maintenance of the 44/52 ratio in human samples except fat and semen (Table 5) indicates rather equal absorption, metabolism, distribution and excretion for the 2 isomers in aqueous-based media. In animal fat, the ratio is low indicating preferential metabolism of PCB 44 by swine, sheep, and chickens; blood ratios were highly variable and PCB 44 was frequently not detected in liver samples. Ratios in catfish livers indicated some preferential metabolism of PCB 44 soon after exposure; however, increased accumulation relative to PCB 52 was observed after long-term exposure. This ob-

Table 5. Ratios of 2,2′3,5′-TCB to 2,2′5,5′-TCB (PCB 44/PCB 52) in various biological samples

Human				
Milk	0.41, 0.43			
Maternal blood	0.44			
Fetal blood	0.43			
Semen	1.50			
Fat	1.0			
Animals fed aroclors	Fat			
Boars (1254 × 6 months)	0.14			
Sows (1242-weaning)	0.03			
Piglets (1242-sows)	0.02			
Gilts (1242 × 3 months)	0.08			
Cockerels (1254 × 20 days)	0.17			
Cockerels (1242 × 8 weeks)	0.03			
Sheep (1242 × 3 months)	0.01			
Channel catfish (A-1242)	50–90 days		112–252 days	
Fat	0.67–0.76		0.65–0.81	
Liver	0.21–0.39		0.54–0.90	
Kidney	–		0.69–0.81	
Rattail liver	Contaminated ocean		Remote ocean	
	0.49 (0.34)[a]		0.07 (0.79)[a]	
Hudson River	Roger's Island		Thompson Island	
	July	August	July	August
Water	0.91 (0.17)[a]	1.21 (2.76)[a]	0.16 (0.09)[a]	1.51 (0.46)[a]
Caddisfly	0.92 (0.25)[a]		1.62 (0.18)[a]	

[a] PCB 87/PCB 52 ratios

servation is consistent with the relative ratios in rattail livers, the low residue fish apparently can preferentially metabolize PCB 44, thus resulting in a lowered ratio to PCB 52. What appeared to be elevated PCB 44 in high residue fish was actually decreased PCB 44 in the other group. The ratio common to commercial PCBs is maintained at high residue levels due to less efficient metabolism of PCB 44.

An additional consideration involves possible anaerobic dechlorination. For example, loss of a *para* chlorine from PCB 87 (2,2′3,4,5′-PCB; water solubility reported as 23 ppm) or PCB 97 or both *para* chlorines from PCB 138 would yield PCB 44. The decreased PCB 87: PCB 52 ratio in high residue fish (Table 5) could indicate *p*-dechlorination of 87 to 44 in sediments. PCB 52 proportions do not change appreciably in the most common acclimated anaerobes in Hudson River sediments, so variations of the ratios in water may indicate participation of PCB 44 in acclimated-sediment dechlorination pathways. Again, the problem of incomplete and inconsistent data (Table 5) complicates interpretations, so that the reverse relationship of PCBs 44 and 87 (and several others) is inconclusive. A similar weaker relationship is seen in Hudson River caddisfly larvae. More dramatic temporal and downstream perterbations in congener ratios in water are seen, but the major products of dechlorination (2-CB and 2,2′-CB) decline rather than increase. Many more, similar data were examined, the main effect being to demonstrate an additional PCB hazard: destruction of large blocks of time in pursuit of intriguing, but elusive, relationships.

3.3 Influence of Residue Composition on Biological Activity

This relationship can be subtle or pronounced and will be addressed in the section on toxicity.

3.4 Variations in Disposition and Composition

It should be apparent that with methodologies as well as PCBs in a state of flux and with inputs of PCB as well as data coming from several sources, estimates, flow patterns and conclusions will seldom agree. It is encouraging that mass balance estimates are within 1 order of magnitude. Again, it should be emphasized that generalizations given here are drawn from reports which have gone to great pains to account for deviations. Some sources of variations will be highlighted briefly in order to maintain perspectives.

Analytical methodologies and quantitation techniques are often cited as critical factors which result in equivocal PCB residue data. Early techniques used marker peaks to estimate total PCB and did not, therefore, account for changes in composition. This effect has been discussed repeatedly with the development of specific congener analysis and need only be mentioned. Specific congener analysis, however, frequently is also limited but totals are more accurate because even peaks lacking standards are estimated to arrive at totals. Long-term data are difficult to interpret since methodologies and accuracies changed even if precision was excellent in the early results. This has led to conclusions that, after the initial impact of regulations, environmental residues of PCB have not declined appreciably since 1978 (Anonymous, 1986; Barros et al., 1984; MacNeill, 1984). A contrary conclusion was reached using data cited as carefully controlled and representing an accurate trend (Swain, 1983). This author sees a continuous decline in PCB loads for 3 species of Great Lakes fish, and projects reaching 2 ppm tolerance levels for coho salmon and lake trout by the end of the century and earlier for the shorter-lived bloater chub.

Variations in data consideration may also have profound influences on generalizations. In Table 2, North American, freshwater sediment PCB load is estimated at 1.4 to 7.1×10^6 Kg most of which is accounted for by the Great Lakes (NAS, 1979). Other estimates are in the range of 0.4–0.9×10^6 Kg for the Great Lakes, but this value would double if only the sediment load of Waukegan Harbor and North Ditch were considered (Eisenreich and Johnson, 1983).

Seasonal flow patterns can also have profound influences on reported PCB loads. It was found that sudden peaks of PCB concentrations in roach from Lake Roxen (Sweden) could be accounted for by increased waterflow carrying contaminated river sediments into the lake (Olsson et al., 1978). The total PCB in Hudson River water at two stable sampling sites did not change appreciably between July and August, 1983, while flow rate nearly doubled; however, the proportions of various lower chlorinated congeners differed dramatically (Bush et al., 1985). These same authors found considerable differences among invertebrate species in level and consistency of residue accumulations.

4 Toxicity of PCBs

4.1 Commercial Mixtures

Acute toxicity is generally not a serious consideration, especially for terrestrial animals (Table 6). It should be noted that rats and fish tend to be less resistant (acutely) to the lower chlorinated PCBs, while birds tend to be less resistant (acutely) to higher chlorinated PCBs. Lethality has been observed in animals which have accumulated high residues and then subjected to stressful conditions (Hansen, 1987).

Sub-acute and chronic effects are many and have been reviewed extensively (Fishbein, 1974; Kimbrough, 1974, 1985; McConnell and Moore, 1979; Safe,

Table 6. Acute toxicities of Aroclors to various animals

Parameter and organism	50% mortality levels for Aroclors					
	1232	1242	1248	1254	1260	1262
LC_{50} (µg/l)[a]						
Grass shrimp (7 day)	–	–	–	3	–	–
Scud (4 day)	–	10	52[b]	2,400[b]	–	–
Crayfish (7 day)	–	30[b]	–	100[b]	–	–
Damselfly (4 day)	–	400	–	200	–	–
Blue gill (4 day)[b]	–	–	278	2,740	–	–
15 day	–	54	76	200	–	–
20 day	–	–	10	140	240	–
Channel catfish						
4 day[b]	–	6,000	12,000	–	–	–
15 day	–	110	130	740	–	–
20 day	–	–	–	300	300	–
Cut-throat trout						
4 day	2,500	5,430	5,750	42,500	60,900	50,000
Dietary (ppm)[c]						
Pheasant	3,200	2,100	1,300	1,100	1,100	1,200
Bobwhite quail	30,000	2,100	1,200	600	750	870
Japanese quail	5,000	5,000	4,800	2,900	2,200	2,300
Oral (mg/kg)						
Rat	4,500	8,700	11,000	10,000	10,000	–

[a] Continuous flow unless specified as b) (Stalling and Mayer, 1972; Mayer et al., 1977)
[b] Static system
[c] 5 days in 2 week birds (NAS, 1979)

Table 7. Summary of common manifestations of PCB exposure in various animals

---

Hepatotoxicity
  Hepatomegaly; bile duct hyperplasia; proliferation of smooth ER
  Focal necrosis; fatty degeneration
  Induction of microsomal enzymes; implicated in hormone imbalances which may explain some
    thyroid, pancreas and reproduction effects
  Depletion of fat soluble vitamins (especially A)
  Porphyria

Immunotoxicity
  Atrophy of lymphoid tissues
  Reduction in circulating leukocytes and lymphocytes
  Suppressed antibody responses
  Enhanced susceptibility to viruses
  Suppression of natural killer cells

Neurotoxicity
  Impaired behavioral responses
  Alterations in catecholamine levels
  Depressed spontaneous motor activity
  Developmental deficits
  Numbness in extremities

Reproduction
  Increased abortion; low birth weights
  Decreased survival and mating success
  Increased length of estrus
  Embryo and fetal mortality
  Gross teratogenic effects (esp. avian species)
  Biochemical, neurological and functional changes following *in utero* exposure (mammals)
  Decreased libido, decreased sperm numbers and motility

Gastrointestinal
  Gastric hyperplasia
  Ulceration and necrosis

Respiratory
  Chronic bronchitis; chronic cough
  Decreased vital capacity

Dermal toxicity
  Chlorance
  Hyperplasia and hyperkeratosis of epithelium
  Edema, alopecia

Mutagenic effects
  Commercial mixtures are weakly mutagenic

Carcinogenic effects
  Preneoplastic changes
  Neoplastic changes (high doses)
  Promotion considered main contribution
  Attenuation of other carcinogens under certain conditions

Table 8. Representative population-modulating responses of various organisms to experimental PCB exposure

Organism	PCB[a]	Lowest effect level[b]	General major effect	Reference
E. coli	A-1242	0.010	Stimulated growth	Keil et al. (1972)
Planktonic algae	Various	0.010–0.100	Decreased growth	Laake (1984)
Centric diatoms			Decreased numbers	Laake (1984)
Dinoflagellates			Increased numbers	Laake (1984)
Diatom *(Cylindrotheca*	A-1242	0.010	No notable effect	Keil et al. (1971)
closteria)	A-1242	0.100	Sharply reduced growth	Keil et al. (1971)
Blue-green Algae	di-Cl, Tri Cl-	0.001–0.100	Inhibited growth	Zullei and Benecke (1978)
(Phormidium)	C-A30	0.050	Inhibited growth	Zullei and Benecke (1978)
	4-Cl, C-A 60	0.100	No effect	Zullei and Benecke (1978)
Green algae	A-1232, 1242	1.0	Transient growth reduction	Hawes et al. (1976)
(Chlorella)	A-1268	1.0	Less effect	Hawes et al. (1976)
(Euglena)	A-1242	10	Depressed growth	Bryand and Olafsson (1978)
Fungus *(Aspergillus*	A-1232, 1254	5	Depressed growth;	Murado et al. (1976)
flavus)	A-1260	25	induction (aldrin epoxidase)	
Protozoan *(Tetra-*	A-1254	0.001	Reduced population	Nimmo et al. (1975)
hymena pyriformis)			growth	
Daphnia magna	A-1248	0.005	Decreased reproduction	Stalling and Mayer (1972)
Grass shrimp	A-1254	0.015	Kills larvae	Roesijadi et al. (1976)
(P. pugio)	A-1254	0.003	Delayed larval development	Roesijadi et al. (1976)
Oyster *(Crassostrea*	A-1254	0.001	No effect	Nimmo et al. (1975)
virginica)	A-1254	0.005	Reduced growth; tissue changes	Nimmo et al. (1975)
Minnow *(Phoxinus)*	C-A50	44–78[c]	Growth stimulation (hormonal)	Bengtsson (1979)
Channel catfish	A-1242	20[c]	Depressed growth in	Hansen et al. (1976)
(Ictaluras)			juveniles; stimulated after 50 g	
Minnow	A-1242	0.015	No spawning	NAS (1979)
(Pimephales)	A-1242	0.005	Reduced spawning	NAS (1979)
	A-1254	0.005	No spawning	NAS (1979)
	A-1254	0.002	Reduced spawning	NAS (1979)
Pinfish *(Lagodon)*	A-1254	0.005	Increased disease susceptibility	Hansen et al. (1971)
Spot *(Leiostomus)*	A-1254	0.005	Increased disease susceptibility	Hansen et al. (1971)
Ring dove *(Streptopelia)*	A-1254	10[c]	No effect 1st clutch; 80% embryo mortality 2nd clutch	Peakall et al. (1972)

[a] A = Aroclor; C = clophen; D-Delor
[b] ppm in culture media, unless noted
[c] ppm in diet
[d] mg/kg. Roughly equivalent to a dietary concentration one order of magnitude greater

Table 8 (continued)

Organism	PCB[a]	Lowest effect level[b]	General major effect	Reference
Laying hens *(Gallus)*	A-1242	20[c]	Embryo mortality	Ax and Hansen (1976)
	A-1248	20[c]	Embryo mortality	Lillie et al. (1974)
	A-1254	20[c]	Embryo mortality	Ax and Hansen (1976)
	PCB 84, 118	20[c]	Embryo mortality	Ax and Hansen (1976)
	PCB 18, 153	20[c]	No effect	Ax and Hansen (1976)
	A.1221, A-1268	20[c]	No effect	Lillie et al. (1974)
Laying hens	D-103	5.0[c]	Decreased bursa; decreased liver vitamin A; increased aniline hydroxylase in chicks	Kosutsky et al. (1979)
Mink *(Mustella)*	A-1254 (residue)	0.64[c]	Reproduction failure; no pup survival	Platonow and Karstad (1973)
	A-1254	2.0[c]	Reproduction failure; no pup survival	Aulerich and Ringer (1977)
	Mixture	3.3[c]	Decreased reproduction	Jensen et al. (1977)
Gray seal (♂) *(Halichoerus)*	A-1254	0.45 (in vitro)	Altered steroid biosynthesis	Freeman and Sangalang (1977)
Swine (♂)	A-1254	100[d]	Altered steroid metabolism	Platonow et al. (1972)
Swine (♀)	A-1242	20[c]	Reduced reproduction; increased stillbirths	Hansen et al. (1975)
Monkeys (♂)	A-1248	5.0[c]	Transient decrease in sperm	Allen et al. (1979)
(Macaca mulatta) (♀)	A-1248	2.5[c]	Fetal deaths; low birth weight behavioral changes in offspring	Allen et al. (1979)
(M. fascicularis) (♀)	A-1248	2[d]	Severe toxicosis	Tryphonas et al. (1984)
	A-1254	0.4[d]	No effect	Tryphonas et al. (1984)
	A-1254	5.0[d]	Severe toxicosis	Tryphonas et al. (1984)

1984; Parkinson and Safe, 1987; Hansen, 1987). Likewise, biochemical effects have been reviewed (Kimbrough et al., 1978; Matthews et al., 1978; Safe, 1984; McKinney et al., 1985), although most studies have focused on the effects of the PCB congeners which are approximate isostereomers of 2,3,7,8-tetrachloro-*p*-dibenzodioxin (TCDD) (Poland et al., 1979) (see Chapter in this book). Residues resulting from PCB exposure may be more (Parkinson et al., 1980) or less (Hansen et al., 1981 a) potent microsomal enzyme inducers than the parent mixture, depending on species involved.

Other effects, with equally broad implications, are documented. Mitochondrial membrane and ATPase effects of PCB mixtures (e.g., Khan and Cutkomp, 1982; Nishihara and Utsumi, 1985) alone are indicative of a broad spectrum of physiological effects. Although less cytotoxic, PCB 153 is a more effective inhibitor of intercellular communication than is PCB 169, a TCDD mimic (Tsushimoto et al., 1983; Rogers-Back and Clark, 1986). This provides an additional mechanism for the cancer promoting and teratogenic effects of PCBs.

PCB 136 as well as certain TCDD mimics can alter brain catecholamine levels in mink (Aulerich et al., 1985) and Aroclors have been shown to have similar effects in rats and doves (Heinz et al., 1980; Seegal et al., 1985). Thermogenesis in brown adipose tissue depends on catecholamine stimulation and the Na/K pump (Horwitz, 1979), processes which can be affected by both TCDD mimics as well as other congeners. A primary target of TCDD is brown adipocytes (Rozman et al., 1986), and various PCBs should interact to modulate the response; only TCDD and very potent mimics produce the characteristic wasting syndrome, but the more subtle effects of PCBs on energy balance and response to cold have not been investigated. Aroclor 1254 has been shown to depress lipid biosynthesis in rat liver (Gamble and Kling, 1976), a further indication of subtle energy-balance effects.

Only a brief outline of documented PCB effects will be presented (Table 7) and these topics may be later expanded, when necessary to clarify organism and/or population effects (Table 8). It must be recognized that there are dramatic species differences in toxic manifestations.

4.2 Patterns of Congener Toxicities

Induction of hepatic microsomal enzymes is one of the earliest and most sensitive parameters to respond to PCBs. PCBs with no *ortho*, 2 *para* and at least 2 *meta* chlorines (i.e., the coplanar PCBs 77, 126 and 169) are very potent mimics of TCDD both in P-450 induction and toxic effects, which are mediated through initial binding to the *Ah* receptor (Poland et al., 1979; Safe, 1984; McKinney et al., 1985; Safe et al., 1985a; Parkinson and Safe, 1987). These congeners, however, are uncommon in commercial PCBs as well as in environmental residues and their mono-*ortho* derivatives (e.g., PCB 105, 118, 156 and 189) may be more important in terms of TCDD-like activity *and* occurrence (Safe, 1984). Certain di-*ortho* derivatives of the 3, 4, 3′, 4′ pattern (e.g., PCB 128, 138, 153, 170, 180) are very significant components of PCB residues (Table 4). PCBs 128, 138 and 170 have reduced TCDD-like effects whereas PDB 153 evokes no TCDD-like responses. This latter PCB resembles phenobarbital (PB) in its mode of induction of P-450 monooxygenases. The coplanar mono- and di-*ortho* derivatives are referred to as mixed inducers (Safe, 1984; Safe et al., 1985a) since they elicit effects similar to co-administration of PB plus MC.

Toxicities other than those associated with the *Ah* locus do not appear to follow strict structure/activity patterns, but there have been few systematic studies. Structural requirements for PCBs yielding metabolites which accumulate in lung parenchyma and/or bronchial mucosa have been determined (Brandt, 1977). Two other reports investigate the effect of structure on PB-type induction activity and elevation of TCDD receptor protein levels – (Denomme et al., 1983, 1986). Moderate chick embryo toxicity has been demonstrated for PCB 52, for example, while PCBs 18 and 153 were inactive and PCBs 84 and 118 were severely toxic but by different mechanisms (Ax and Hansen, 1975; Ax et al.,1976; Hansen, 1987). Toxic responses unrelated to *Ah* locus effects and less intense than those for Aroclor 1254 have been reported for, e.g., PCBs 4, 28, 31, 49, 52, 84, 95, 110, 136 and 153 (Vos and Notenboom-Ram, 1974; Hansen, 1987; Torok, 1976).

4.3 Toxicities Related to PCB Metabolites

PCBs are generally metabolized to hydroxylated products directly or through arene oxide intermediates (Sundström et al., 1976; Sipes, 1987). This metabolism is facilitated by the presence of vicinal hydrogens and is reflected in the composition of PCB residues (Jensen and Sundström, 1974). Phenolic metabolites are more acutely toxic than parent PCBs (Yamamoto and Yoshimura, 1973; Standnicki and Allen, 1979) and more readily metabolized congeners tend to react (covalently) more readily with cellular macromolecules such as protein, DNA and RNA (Morales and Matthews, 1979; Brandt et al., 1981; Safe, 1984; Hansen, 1987).

This binding may be responsible for inhibition of some microsomal enzyme catalyzed reactions such as hexabarbital hydroxylase (Schmoldt et al., 1977; Shull et al., 1982) and the monooxygenases which metabolize other PCBs (Hansen, 1987). Accumulation of an acid lactone metabolite of PCB 18 in cultures of aerobic marine bacteria was thought to inhibit further metabolism, but accumulation in the marine environment would probably not be significant (Carey and Harvey, 1978). Likewise, phenolic metabolites are rapidly conjugated and probably of little food chain significance; however, phenolic metabolites generated *in situ* can inhibit various P-450 monooxygenase reactions (Schmoldt et al., 1977).

Methyl sulfide and methyl sulfone metabolites, on the other hand, have been demonstrated to be present at significant levels in mammalian tissues (Mio et al., 1976; Jensen and Jansson, 1979; Haraguchi et al., 1984). Gastrointestinal flora are necessary for formation of these sulfur metabolites and they have remarkable tissue-specific affinities (Brandt et al., 1985). There is no evidence for formation of sulfur containing metabolites by fish, but a non-polar metabolite of PCB 18 is stored at high levels by algae (*Oedegonium*) and snails (*Physa*) in a model aquatic ecosystem (Metcalf et al., 1975). Mass spectral data are consistent with a sulfur containing metabolite (R. L. Metcalf, personal communication). Methyl sulfonyl metabolites of chlorobenzenes are more potent than parent compounds as PB-type inducers (Kimura et al., 1983; 1985; Kato et al., 1986). The absence of both these PCB metabolites and this type of PCB induction in fish suggests a possible relationship, even if phenobarbital is only weakly effective in fish.

5 Experimental Effects on Biota

There is an abundance of data on various effects of PCBs on various life forms and the preponderance of information is concerned with induction of microsomal oxidases. Microsomal cytochrome P-450 induction is a sensitive monitor for biological activity of PCBs and qualitative as well as quantitative differences can be used to predict potencies and types of effects.

Caution cannot be overstressed when relationships established for rats are applied to other species, especially *in vitro*. The trend is to apply rodent methods and conditions to other orders, classes and phyla. Swine pH and temperature optima are higher than those in the "standard rodent recipe". The pH optimum for avian microsomes is generally 6.8–7.2 (Grossman and Khan, 1981; Shackelford and Khan, 1981). Temperature optima for birds are a surprising 42–45 °C, those for invertebrates are about 20–30 °C and fish are variable, but generally below 25 °C. Small variations from several optimum preparative and assay conditions can compromise activity to the extent that casual species and/or treatment comparisons may be meaningless (Hansen, 1983; James, 1986).

Induction of the 3-MC type can be demonstrated in fungi (Murado et al., 1976). In fish, MC-type induction results from PCB exposure, but PB-type is not readily apparent (Bend et al., 1977; Stegeman, 1978; Forlin, 1980; Lech and Petersen, 1983). Higher enzyme activities are found in deep ocean fish from more contaminated areas with higher liver PCB residues than in the same species from a more remote area (Stegeman et al., 1986). The elevated enzyme activity in fish liver was attributed to induction by PCB; on the other hand, elevated mixed function oxidase activity in water snakes from areas with high organochlorine contamination was due to inherent species differences which permitted survival, while species with low oxidase activity and/or oviparous reproduction were eradicated (Stafford et al., 1976). Birds and mammals show both types as well as mixed induction by commercial mixtures and certain congeners (Safe, 1984). Avian species differ from mammals with respect to the monooxygenase enzymes induced by MC- and PB-type inducers (Jones et al., 1985).

Adequate evidence exists for considering factors other than *Ah* locus effects in evaluating PCB exposure. In chick embryos, induction of enzyme activity by TCDD and PCBs 77 and 169 is not correlated with lethality (Rifkind et al., 1985). If PCBs 84 and 118 are included in hen diets at 20 ppm, microsomal monooxygenase enzyme activity is induced by 118 and depressed by 84, but both result in 100% embryo mortality by 4–5 weeks (Ax et al., 1976; Hansen, 1987). Aminopyrine N-demethylase activity, shown to be associated with TCDD induction in the avian (Jones et al., 1985), correlated negatively with PCB residues in colony-collected herring gull embryos (Boersma et al., 1986). When comparing the highly PCB-sensitive mink to the related but less sensitive ferret, microsomal enzyme induction was inversely related to toxic manifestations (Shull et al., 1982). It has further been demonstrated that certain congeners with no *Ah* locus activity elevate the cytosolic *Ah* receptor protein, increasing the potential activity of TCDD mimics (Denomme et al., 1986).

Other effects such as outlined in Table 7 are manifest in varying degrees depending on PCB, species, sex and developmental stage. Some examples of responses especially relevant to environmental toxicology (i.e., reproduction, disease resistance, ability to compete for available niches) are presented in Table 8. The list is merely representative and by no means complete.

Plants from algae and yeast to soybeans readily absorb lower chlorinated biphenyls from aqueous media (Cole et al., 1979; Suzuki et al., 1977). Simple plants tend to be more susceptible to lower chlorinated biphenyls, but the effect may be stimulative as well as inhibitory. In addition, some types of plants are more sensitive than others and some effects are transient; thus, population profiles are likely to shift at levels greater than 5–10 ppb which are found in surface waters of the North Atlantic (NAS, 1979) and some of the Great Lakes (Eisenreich and

Johnson, 1983). Alterations such as these can indirectly affect entire trophic systems even if total phytoplankton are not reduced by PCBs (Fisher, 1975; O'Connors et al., 1978).

Dietary exposures up to 20 ppm for less than a lifetime are not unrealistic experimentally, and these levels can have important effects on fish, birds and mammals (Table 8). Again, demonstrated different effects and variable sensitivities can alter terrestrial as well as aquatic and marine population profiles.

6 Population and Ecological Changes Attributed to PCBs

The selective toxicity of PCBs to phytoplankton (see Table 8) was confirmed in mixed cultures at concentrations of 10 ppb Aroclor 1254 (Fisher, 1975). Although the total photosynthetic capacity of contaminated marine communities may not be diminished, the species profile most certainly would change. Concentrations as low as 1 ppb Aroclor 1254 suppressed large phytoplankton while not affecting various species below 9 μm equivalent spherical diameter (O'Connors et al., 1978). The growth suppression was transient, however recovery took at least 3 days and concentrations of 5 or 10 ppb had more pronounced effects. It was proposed that the shift to phytoplankton communities dominated by smaller forms would result in longer food chains favoring gelatinous predators rather than harvestable marine fish.

A polluted system long adapted in manners similar to the phytoplankton example above is a poor subject for retrospective elucidation of environmental effects of PCBs. Failure to demonstrate effects on invertebrate and fish diversity in adapted ecosystems lacking top predators is certainly inconclusive (Laake, 1984). Historical trends must be included with current data, thoughtful experimentation and insight. Such a combination showed that the distribution of aquatic snakes in three distinct ecosystems in central Texas was related to pesticide pollution and to large variations in detoxifying enzymes (Stafford et al., 1976). *N. rhombifera* had significantly lower mixed function oxidase activities and was the only species not collected from the highly contaminated cotton field runoff ecosystem (Table 9); this species had been collected prior to 1950 but was presumed eradicated due to this low enzyme activity. Only the shed skin from one oviparous snake was found in 1971 however various oviparous species appeared to be re-establishing in 1974–1975 and this coincided with declining DDE residues (Fleet and Plapp, 1978). Habitat rather than PCB appeared responsible for the limited species in the highway site (Stafford et al., 1976). Mixed function oxidase enzyme activities did not vary significantly from site to site for the same species indicating that enzyme induction and/or focal selection for higher enzyme activity were not responsible for the species differences in metabolism.

In the previous example it appears that DDT and DDE were primarily responsible for the transient decline in oviparous snakes and the eradication of *N. rhombifera*. The effect could be intensified by development of resistance among prey species as was demonstrated in a similar agro-ecosystem (Rosato and Ferguson,

Table 9. Mixed function oxidases[a] and chlorinated hydrocarbon residues[b] in water snakes (*Natrix* spp.) and copperheads and water moccasins (*Agkistrodon* spp.) collected in 1972–1973 from ecosystems variously contaminated (Stafford et al., 1976; Fleet and Plapp, 1978)

Species	Pesticide			Non-pesticide			Hiway ditch		
	mfo	DDE	PCB	mfo	DDE	PCB	mfo	DDE	PCB
N. rhombifera	NA[c]	NA	NA	8.9	2.3	BD	6.1	16.4	47.0
N. eryrthogaster	12.2	787	BD[d]	13.6	2.4	BD	14.8	15.0	37.2
(1975–5)	–	179	–	–	–	–	–	–	–
N. fasciata	10.6	671	BD	NA	NA	NA	NA	NA	NA
(1974–5)	–	212	–	–	–	–	–	–	–
A. piscavorus	38.6	308	BD	NA	NA	NA	NA	NA	NA
(1974–5)	–	216	–	–	–	–	–	–	–
A. contortrix	47.5	14.4	BD	43.6	1.7	BD	NA	NA	NA
(1974–5)	–	114	–	–	–	–	–	–	–

[a] Percent testosterone metabolized by liver homogenates. Significant sex difference for *N. rhombifera* only
[b] PPM in fat (PCB as Aroclor 1254)
[c] NA = speciemens not collected
[d] BD = Below detection limit

1968). In other cases such as reproduction failure in mink, pesticides were first suspected but Aroclor 1254 residues were ultimately shown to be the most potent factor (Aulerich and Ringer, 1977). Direct relationships between reproduction problems and chlorinated aromatics in brown bats have led to a postulated causal relationship. PCB, quantitated as Aroclor 1260, were more predominant as redidues than DDE (Clark and Lamont, 1976); subsequent feeding with Aroclor 1260 and more detailed examination of the effects of age, indicated that maternal age was a more important factor in bat stillbirths than was Aroclor 1260 (Clark, 1978). Nevertheless, Aroclor 1260 is less embryotoxic than Aroclor 1254 in other species and Aroclor 1254 residues tend to resemble Aroclor 1260 after passing through various trophic levels; therefore, the influence of PCB can not be unequivocally dismissed as an etiologic agent in bat toxicosis even though it may not be the major contributor to this toxicosis.

More convincing evidence for associating PCB with reproduction problems was presented for colonies of terns (Common terns, *Sterna hirundo* and Roseate terns, *S. dougallii*) on Great Gull Island in the eastern end of Long Island Sound (Hays and Risebrough, 1971; 1972). Abnormal young terns (feather loss; eye, bill, and foot deformities) increased from 0.1% in 1969 to 1.3% in 1970; in addition, feather loss from other tern colonies in Long Island Sound were reported for the first time in 1970. No significant microbial pathogens were found. Mercury levels did not appear abnormally high and DDT/DDE levels were lower than in many areas; however, PCB (resembling Aroclor 1254) levels in fish caught by terns were quite high (10–176 ppm, mean of 50 ppm in lipids). Concentrations of DDE in breast muscle of young abnormal terns were 0.5–9.0 ppm freshweight, while PCB concentrations varied from 5–140 ppm in breast muscle (Hays and Risebrough, 1972). The possible role of polychlorinated dibenzofurans and dibenzo-p-dioxins in this toxicosis were discussed; however, since PCBs manufactured in North America, contain low levels of these highly toxic contaminants, it must be concluded that PCBs were primarily responsible for these effects. The bill deformity had previously been demonstrated in a domestic chick developing in an egg injected with Aroclor 1242. We have observed leg, liver and down abnormalities in chicks from hens fed 20 ppm of the highly embryotoxic PCB 118, a major component of Aroclor 1254 (Ax et al., 1976); another group of chicks in the same study had crossed beaks and encephaloceles but were not reported due to incubator malfunctions (W. L. Miller and L. G. Hansen, unpublished results). Similar effects, including crossed beaks, were observed by Cecil et al. (1974) in chicks fed dietary levels of 10–20 ppm Aroclors. Thus, there is very strong evidence that certain PCBs are teratogenic in avian species and the concentrations in muscle tissue of abnormal tern embryos were certainly high enough to indicate a causal relationship.

Chlorinated hydrocarbon pesticides in Great Lakes fish were also dismissed as the cause of reproductive failure in mink and Aroclor 1254 and related PCB residues were established as the major agents (Aulerich and Ringer, 1977; Jensen et al., 1977). Diminished seal populations are believed to be related to PCB contamination and the definitive studies with mink, another fish-eating mammal with delayed implantation, strengthened the correlation. Pathological changes in seal uteri which are associated with high PCB levels (Helle et al., 1976a; 1976b), are not observed in mink thus illustrating some species differences in these effects. The higher PCB levels in non-pregnant female seals with uterine pathology may merely reflect loss of an elimination route (transfer to fetus), but non-pregnant females with normal uteri did not differ significantly from pregnant females in PCB levels.

Stable populations of grey seals (*Halichoeaus grypus*) and harbor seals (*Phoca vitulina*) from Norwegian coastal waters contained much lower PCB and total DDT residues than those from areas where the populations are threatened (Ofstad and Martinsen, 1983). Furthermore, in the Wadden Sea, there is a direct correlation between increasing reproductive success and decreasing PCB concentration as the seal populations become further removed from the mouth of the Rhine River (Dutch Wadden Sea); dilution of the contaminated river water by the North Sea results in lower PCB residues in stable Schleswig-Holstein and Danish seal populations (Reijnders, 1980). Threatened Dutch seal populations contain higher PCB/DDE ratios than threatened populations in the Baltic (Table 10). Uterine pathology observed in Baltic seals (Helle et al. 1976b) is not seen in Wadden Sea seals where the reproductive failure is attributed to failed implantation and/or early abortion (Reijnders, 1980). Certainly both DDE and PCBs influence hormonal balance, PCBs are fetotoxic and at least PCBs are immunosuppressive. The interactions of these high residues of PCBs and related contaminants appear to compromise seal reproduction in various areas by diverse biological processes which have not hitherto been determined.

Earlier studies with California sea lions (*Zalophus californianus californianus*) revealed 7.6 times greater DDE residues and 4.4 times greater PCB residues in

Table 10. Comparative DDT and PCB concentration (ppm) in blubber from various seal populations

	Grey seal		Harbor seal		References
	DDT	PCB	DDT	PCB	
Threatened populations					
Gulf of Bothnia	210 ± 28	100 ± 18	–	–	Helle (1976)
Baltic Proper	420 ± 53	140 ± 17	–	–	Helle (1976)
Dutch Wadden Sea	–	–	47.3	701	Reijnders (1980)
Stable populations					
East Wadden Sea	–	–	8.5	76.4	Reijnders (1980)
Norwegian Coast	0.4–8.2	1.9–33	1.1–7.7	1.6–30	Ofstad and Martinsen (1983)
Swedish W. Coast	–	–	46 ± 11	60 ± 12	Jensen and Jansson (1979)

blubber from females delivering premature pups (DeLong et al., 1973). These large differences in residues were attributed to wintering in different areas (De Long et al., 1973) and it was established that the younger females tended not to migrate far from the more polluted Channel Islands than did the older females wintering further south (Gilmartin et al., 1976). The high levels of organochlorine contaminants were, as opposed to the conclusions reached with bats (Clark, 1978), nevertheless, associated with premature parturition through immune suppression and interactions with other contaminants (Gilmartin et al., 1976).

Interactions with other toxicants have been demonstrated experimentally and implicated as enhancing PCB effects on various populations. PCBs can enhance Cd accumulation in mink kidneys (Olsson et al., 1979) and a similar parallel trend between lipid PCB and renal Cd was seen in swine exposed to sewage sludge fertilized soild (Hansen et al., 1981), although the confounding factors in the latter study were considerable. Certain PCBs can increase the formation of detoxified aflatoxin B_1 metabolites and inhibit the carcinogenic and mutagenic potential of this procarcinogen (Halvorson et al., 1985; Shelton et al., 1984). On the other hand, the rate of formation of the carcinogenic 7,8- dihydrodiol metabolite of benzo(a)pyrene by rat liver microsomes is markedly increased by pretreatment of the rats with 3-MC type PCB inducers; in contrast microsomes from rats pretreated with mixed-type PCB inducers increased the rate of formation of all benzo(a)pyrene metabolites (Haake et al., 1985).

English sole (*Parophrys vetulus*) from 11 sites in Puget Sound showed a high correlation between total liver lesions and biliary concentrations of polycyclic aromatic hydrocarbon (PAH) metabolites (Krahn et al., 1986); however, the highest incidence of neoplasms and foci of cellular alteration was in the fish from the Duwamish Waterway, which is an area which contains moderate levels of PAH and relatively high levels of PCBs (330 ppb and 10X the next lowest level) in the sediment. Since fish do not respond to PB type induction, the mixed type PCB inducers would resemble 3-MC in their mode of induction and increase the rate of formation of the most carcenogenic PAH metabolites. Therefore, the presence of PCB may partially explain the disproportionate response to PAH.

The drastic decline in the isolated beluga whale (*Delphinapterus leucas*) population in the St. Lawrence estuary is also believed to be associated with the interactions among several contaminants, including PCB. This population did not recover as did the arctic populations when hunting was discontinued and various pathological changes have been associated with high levels of PAH and organochlorine pollutants (Martineau et al., 1985; Masse et al., 1986). Blubber PCB concentrations were 3–90 fold higher than those in a beluga whale from the Baltic (Harms et al., 1978). A metastisized urinary bladder carcinoma indicated possible PCB/PAH interaction as previously noted with fish. There has been considerable evidence for increased disease susceptibility associated with PCB as in estuarine fish and the California sea lions. Other documented subtle effects of PCB on mammalian reproduction must exert some effect since several beached whales had blubber residues in excess of 100 ppm PCB. In addition to septicemia and pneumonia, a juvenile whale with 576 ppm blubber PCB had a perferated gastric ulcer, chronic hepatitis and generalized dermatosis which are conditions consistent with experimental PCB toxicosis in other mammals.

Agents in addition to PCB are undoubtedly contributing to the St. Lawrence beluga whale decline. Major efforts in research, regulation and public information have attenuated the rate of chlorinated hydrocarbon contamination of the global environment, but the beluga whale decline is distressing because the data are so recent. Improvement in conditions and recovery of some populations seems indicated, but continued vigilance, increased understanding and careful management of sensitive balances must be carefully maintained.

7 Apparent Trends in PCB Disposition and Biotic Effects

Analysis of aquatic and marine sediments indicate that detectable environmental dispersion of PCBs began in the mid-1930's, became significant in the 1950's and peaked in the early 1970's (Eisenreich and Johnson, 1983; Sugiura et al., 1986). It is conceivable that heightened concern in the early 1970's encouraged a pulse of discharge in anticipation of stricter regulations. The intense awareness of chlorinated aromatics in the 1960's and regulations in the 1960's and early 1970's, however, appear to be paying dividends.

Mean concentrations of total DDT-derived residues in human fat and milk peaked in the late 1960's and declined during the early 1970's; the frequency of detectable residues in analytical surveys has remained near 100%, however the maximum levels have generally declined (Westöö, 1974; Kutz et al., 1979). DDT residues in ecosystems also declined during the 1970's (e.g., Fleet and Plapp, 1978). In contrast, during the period (1967–1972) when DDT residues were declining in Sweden, PCB residues were increasing (Westöö, 1974).

The rate of decline in environmental (including human) PCB residue concentrations between the early and late 1970's has significantly decreased and has led to some concern that "ambient" levels of PCBs have stabilized due to the huge environmental reservoirs of this pollutant (Anonymous, 1986; Barros et al., 1984). The frequency of detectable residues in all classes of vertebrate wildlife is increasing, but mean and maximum residue levels are declining in highly contaminated areas and geographical distinctions are much less clear than in the 1970's (Eisler, 1986). Various forces of dispersion, particularly atmospheric transport, appear to be contributing to the entropic fate of mobile PCB reservoirs.

It appears that PCB residues in Lake Michigan bloater chubs have declined since 1972, whereas residues in lake trout and coho salmon peaked in 1974; these differences were probably due to food chain and life cycle factors (Swain 1983). During this time, however, the PCB burden of Lake Superior has been increasing (Eisenreich and Johnson, 1983; Hallett, 1984).

In addition to dilution by dispersion, it does appear that some PCB degradation is taking place, but rather slowly. Anaerobic bacteria acclimated to relatively high sediment PCB concentrations can slowly dechlorinate PCBs with half-lives measured in years (Brown et al., 1985). At the same time, lower chlorinated congeners appear to be disappearing so that residues that once resembled Aroclor 1242, now resemble the more highly chlorinated Aroclor 1254 (Eisler, 1986). Some PCB metabolism, particularly of the lower chlorinated congeners, occurs at all phylogenetic levels; continued release of PCBs, even at low levels (Barros et al., 1984; Eisler, 1986) and/or slow dechlorination (Brown et al., 1985) may provide adequate levels of labile congeners to render changes in composition too subtle for detection by many analytical techniques.

The trends in ecological effects are contradictory at times. For example, in the Great Lakes, the populations of herring gulls and double-crested cormorants are recovering (Boersma et al., 1986; Eisler, 1986); however, the beluga whale population in the St. Lawrence estuary which drains the Great Lakes is not recovering (Martineau et al., 1985; Masse et al., 1986).

8 Trends in Research

After a decade of intensive, but apparently unfocused PCB research, a limited explanation of the mechanism of toxic action of coplanar and monoortho coplanar PCBs has been proposed, namely the requirement for an initial interaction with the 2,3,7,8- TCDD (or Ah) receptor protein (Safe, 1984). However, it is not clear that this mechanism is operative for all the observed activities elicited by PBCs. Other toxic effects, including interactions with other toxicants, may proceed via multiple mechanistic pathways which have not yet been delineated. Furthermore, retrospective evaluation of some studies with PCBs has frequently revealed flawed conclusions only apparent after further work, such as establishing the extent and influence of dibenzofuran contamination (Voss et al., 1970; Nagayama et al., 1976; Goldstein et al., 1978). Nevertheless, responsible and effective evaluation and management of PCB problems is hampered by dismissing ambigous effects or attributing them to contaminants (Hansen, 1987). PCB toxicoses are so complex that an entire spectrum of effects is never seen in an individual or in every species. Recent trends are expanding the spectrum of PCB effects and acknowledging at least a modulating role for "nontoxic" congeners (McKinney et al., 1985; Safe et al., 1985a).

9 Summary

PCBs are complex mixtures which have been identified throughout the global environment; these pollutants are frequently accompanied by high, and more rapidly declining residues of other chlorinated hydrocarbons. Various PCB mixtures have different qualitative and quantitative biological activities, and these difference can be attributed to the relative concentrations of isomers and congeners present in these mixtures. Environmental residues are more complex than the commercial mixtures and their action is complicated by interactions with other contaminants, hormones and the immune system (thus with microbes) and by interactions with various physiological and biochemical processes in ways not thoroughly understood.

Some adaptation of ecosystems has occurred and this complicates attempts to define PCB effects. Focusing on only the more concise relationships further obscures environmental evaluation. In spite of the overwhelming complexity and obscurring effects, it is clear that PCBs have a detrimental effect on various ecosystems.

PCB input into the environment has declined but not ceased. Variations in sampling and analytical techniques make it difficult to accurately define trends; however, it appears that the PCBs in many mobile environmental reservoirs are declining or shifting to less mobil sinks such as ocean sediments. Atmospheric movement, however, is increasing PCB loads in less contaminated areas (e.g., Lake Superior) while PCB loads in the more contaminated areas are declining. Careful management and observation are necessary to minimize population (including human) and ecosystem effects.

10 References

de Alencastro LF, Prelaz V, Tarradellas J (1984) Bull. Environ. Contam. Toxicol. *33*:270
Allen JR, Barsotti DA, Lambrect LK, van Miller JP (1979) Ann. N.Y. Acad. Sci. *320*:419
Anonymous (1986) Chem. Engin. News *64*:24
Aulerich RJ, Bursian SJ, Breslin WJ, Olson BA, Ringer RK (1985) J. Toxicol. Environ. Hlth. *15*:63
Aulerich RJ, Ringer RK (1977) Arch. Environ. Contam. Toxicol. *5*:279
Ax RL, Hansen LG (1975) Poult. Sci. *54*:895
Ax RL, Miller WL, Hansen LG (1976) J. Anim. Sci. *42*:1365
Baker EL, Landrigan PJ, Glueck CJ, Zack MM, Liddle JA, Burse VM, Housworth WJ, Needham L (1980) Am. J. Epidemiol. *112*:553

Ballschmiter K, Zell M (1980) Fresenius Z. Anal. Chem. *302*:20

Barros MC, Konemann H, Visser R (eds) (1984) PCB Seminar Proceedings, 410 pp., Ministry of Housing, Physical Planning and Environment, The Hague

Bedard DL, Unterman R, Bopp LH, Brennan MJ, Haberl ML, Johnson C (1986) Appl. Environ. Microbiol. *51*:761

Bend JR, James MO, Dansette PM (1977) Ann. N.Y. Acad. Sci. *298*:505

Bengtsson BE (1979) Ambio *8*:169

Boersma DC, Ellenton JA, Yagminas A (1986) Environ. Toxicol. Chem. *5*:309

Brandt I (1977) Acta Pharmacol. Toxicol. *40* (Suppl. II):1

Brandt I, Lund J, Bergman A, Klasson-Wehler E, Poellinger L, Gustafsson JA (1985) Drug Metab. Dispos. *13*:490

Brandt I, Mohammed A, Slanina P (1981) Toxicol. *21*:317

Brown JF, Wagner RE, Bedard DL, Brennan MJ, Carnahan JC, May RJ, Tofflemire TJ (1985) Preprint, ACS Div. Environ. Chem. *25*:35

Bryand AM, Olafsson PG (1978) Bull. Environ. Contam. Toxicol. *19*:374

Burkhard LP, Andren AW, Armstrong DE (1985a) Environ. Sci. Technol.*19*:500

Burkhard LP, Armstrong DE, Andren AW (1985b) Environ. Sci. Technol. *19*:590

Burse VW, Needham LL, Lapeza CR, Karver MP, Liddle JA, Bayse DD (1983) J. Assoc. Offic. Anal. Chem. *66*:456

Bush B, Shane LA, Wilson LR, Barnard EL, Barnes D (1986) Arch. Environ. Contam. Toxicol. *15*:285

Bush B, Simpson KW, Shane L, Koblintz RR (1985) Bull. Environ. Contamin. Toxicol. *34*:96

Bush B, Tumasonis CE, Baker FD (1974) Arch. Environ. Contam. Toxicol. *2*:195

Carey AE, Harvey GR (1978) Bull. Environ. Contam. Toxicol. *20*:527

Cecil HC, Bitman J, Lillie RJ, Fries GF, Verrett J (1974) Bull. Environ. Contamin. Toxicol. *11*:489

Clark DR Jr (1978) Bull. Environ. Contam. Toxicol. *19*:707

Clark DR Jr, Lamont TG (1976) Bull. Environ. Contam. Toxicol. *15*:1

Cole MA, Reichart PB, Button DR (1979) Bull. Environ. Contam. Toxicol. *23*:44

Collier JL, Hurley SS, Welborn ME, Hansen LG (1976) Bull. Environ. Contam. Toxicol. *16*:182

DeLong RL, Gilmartin WG, Simpson JG (1973) Science *181*:1168

Denomme MA, Bandiera S, Lambert I, Copp L, Safe L, Safe S (1983) Biochem. Pharmacol. *32*:2955

Denomme MA, Leece B, Li A, Towner R, Safe S (1986) Biochem. Pharmacol. *35*:277

D'ltri FM, Kamrin MA (1983) PCBs: Human and Environmental Hazards, Butterworth, Boston, 443 pp

Eisenreich SJ, Johnson TC (1983) In: D'ltri and Kamrin, p. 49

Eisler R (1986) U.S. Fish Wildl. Serv. Biol. Rep. *85*(1.7):72 pp

Fishbein L (1974) Ann. Rev. Pharmacol. *14*:139

Fisher NS (1975) Science *189*:463

Fleet RR, Plapp FW Jr (1978) Bull. Environ. Contam. Toxicol. *19*:383

Freeman HC, Sangalang GB (1977) Arch. Environ. Contam. Toxicol. *5*:369

Friend M, Trainer DO (1970) Science *170*:1314

Fries GF (1982) J. Environ. Qual. *11*:14

Forlin L (1980) Toxicol. Appl. Pharmacol. *54*:420

Gamble W, Kling D (1976) Fed. Proc. *35*:1625

Gilmartin WG, DeLong RL, Smith AW, Sweeney JC, DeLappe BW, Risebrough RW, Griner LA, Dailey MD, Peakall DB (1976) J. Wildlife Dis. *12*:104

Goldstein JA, Haas JR, Linko P, Harvan DJ (1978) Drug Metab. Dispos. *6*:258

Grossman JC, Khan MAQ (1981) Comp. Biochem. Physiol. *63*C:251

Haake J, Merrill J, Safe S (1985) Can. J. Physiol. Pharmacol. *63*:1096 Toxicologist. *5*:35

Hallett DJ (1984) In: Barros et al., p. 80

Halvorson MR, Robertson L, Safe S, Phillips TD (1985) Appl. Environ. Microb. *49*:882

Hansen DJ, Parrish PR, Lowe JI, Wilson AJ Jr, Wilson PD (1971) Bull. Environ. Contamin. Toxicol. *6*:113

Hansen LG (1983) Neurotoxicol. *4*:97

Hansen LG (1987) Rev. Environ. Toxicol. *in press*

Hansen LG, Byerly CS, Metcalf RL, Bevill RF (1975) Am. J. Vet. Res. *36*:23

Hansen LG, Chaney RL (1984) Rev. Environ. Toxicol. *1*:103

Hansen LG, Strik JJTWA, Koeman JH, Kan CA (1981 a) Toxicol. *21*:203

Hansen LG, Tuinstra LGMTh, Kan CA, Strik JJTWA, Koeman JH (1983) J. Agr. Fd. Chem. *31*:254

Hansen LG, Washko PW, Tuinstra LGMTh, Dorn SB, Hinesly TD (1981 b) J. Agr. Fd. Chem. *29*:1012

Hansen LG, Wiekhorst WB, Simon J (1976) J. Fish. Res. Bd. Canada, *33*:1343

Haraguchi H, Kuroki H, Masuda Y, Shigematsu N (1984) Food Cosmet. Toxicol. *22*:283

Harms U, Drescher HE, Huschenbeth E (1978) Meeresforsch. *26*:153

Hawes ML, Kricher JC, Urey JC (1976) Bull. Environ. Contam. Toxicol. *15*:14

Hays H, Risebrough RW (1971) Nat. Hist. *80*:39

Hays H, Risebrough RW (1972) Auk. *89*:19

Heinz GH, Hill EF, Contrera JF (1980) Toxicol. Appl. Pharmacol. *53*:75

Helle E, Olsson M, Jensen S (1976a) Ambio. *5*:188

Helle E, Olsson M, Jensen S (1976b) Ambio. *5*:261

Horn EG, Hetling LJ, Tofflemire TJ (1979) Ann. N.Y. Acad. Sci. *320*:591

Horwitz BA (1979) Fed. Proc. *38*:2170

Humphrey HEB (1984) In: D'ltri and Kamrin, p. 229

Hutzinger O, Safe S, Zitko V (1974) The Chemistry of PCB's, CRC Press, Cleveland

Iwata Y, Gunther FA (1976) Arch. Environ. Contam. Toxicol. *4*:44

James MO (1986) Vet. Human Toxicol. *28* (Suppl. 1):2

Jensen AA (1984) p. 81 in Barros et al.

Jensen S, Jansson B (1979) Ann. N.Y. Acad. Sci. *320*:436

Jensen S, Johnels AG, Olsson M, Otterlind G (1969) Nature *224*:247

Jensen S, Kihlstrom JE, Olsson M, Lundberg C, Orberg J (1977) Ambio. *6*:239

Jensen S, Sundström G (1974) Ambio. *3*:70

Jones D, Sawyer TW, Rosanoff K, Safe S (1985) Toxicologist. *5*:35

Kato Y, Kogure T, Sato M, Murata T, Kimura R (1986) Toxicol. Appl. Pharmacol. *82*:505

Keil JE, Graber CD, Priester LE, Sandifer SH (1972) Environ. Hlth. Persp. *1*:175

Keil JE, Priester LE, Sandifer SH (1971) Bull. Environ. Contam. Toxicol. *6*:156

Khan HM, Cutkomp LK (1982) Bull. Contam. Toxicol. *29*:577

Kimbrough RD, Buckley J, Fishbein L, Glamm G, Kasza L, Marcus W, Shibko S, Teske R (1978) Environ. Hlth. Persp. *24*:173

Kimbrough RD (1974) Crit. Rev. Toxicol. *2*:445

Kimbrough RD (1985) Environ. Hlth. Persp. *59*:99

Kimura R, Kawai M, Kato Y, Sato M, Aimoto T, Murata T (1985) Toxicol. Appl. Pharmacol. *78*:300

Kimura R, Kawai M, Sato M, Aimoto T, Murata T (1983) Toxicol. Appl. Pharmacol. *67*:338

Klein H (1984) In: Barros et al., p. 66

Koeman JH, ten Noever de Brauw MC, de Vos RH (1969) Nature *221*:1126

Kosutsky J, Adamec O, Bobakova E, Sarnikova B (1979) Zivocisna Vyroba. *24*:659

Krahn MM, Rhodes LD, Myers MS, Moore LK, MacLeod WD Jr, Malins DC (1986) Arch. Environ. Contam. Toxicol. *15*:61

Kuroki H, Masuda Y (1977) Chemosphere *6*:469

Kutz FW, Strassman SC, Sperling JF (1979) Ann. N.Y. Acad. Sci. *320*:60

Laake M (1984) In: Barros et al., p. 119

Lech JJ, Peterson RE (1983) In: D'ltri and Kamrin, p. 187

Lillie RJ, Cecil HC, Bitman J, Fries GF (1974) Poult. Sci. *53*:726

MacNeil JW (1984) In: Barros et al., p. 16

Martineau D, Lagace A, Masse R, Morin M, Beland P (1985) Can. Vet. J. *26*:297

Masse R, Martineau D, Tremblay L, Beland P (1986) Arch. Environ. Contam. Toxicol.

Matthews H, Fries G, Gardner A, Garthoff L, Goldstein J, Ku Y, Moore J (1978) Environ. Hlth. Persp. *24*:147

Mayer FL, Mehrle PM, Sanders HO (1977) Arch. Environ. Contam. Toxicol. *5*:501

McConnell EE, Moore JA (1979) Ann. N.Y. Acad. Sci. *320*:138

McKinney JD, Chae K, McConnell EE, Birnbaum LS (1985) Environ. Hlth. Persp. *60*:57

Mes J, Davies DJ, Lau PY (1980) Chemosphere *9*:763

Metcalf RL, Sanborn JR, Lu P-Y, Nye D (1975) Arch. Environ. Contam. Toxicol. *3*:151

Mio T, Sumino K, Mizutani T (1976) Chem. Pharm. Bull. *24*:1958

Morales NM, Matthews HB (1979) Chem. Biol. Interact. 27:99

Mullin MD, Pochini CM, McCrindle S, Romkes M, Safe SH, Safe LM (1984) Environ. Sci. Technol. *18*:468

Murado MA, Tejedor MC, Baluja G (1976) Bull. Environ. Contam. Toxicol. *15*:768

Nagayama J, Kuratsune M, Masuda Y (1976) Bull. Environ. Contam. Toxicol. *15*:9

National Academy of Sciences (NAS) (1979) Polychlorinated biphenyls. Nat. Res. Council, NAS, Washington, DC, 182 pp.

Nimmo DR, Hansen DJ, Couch JA, Cooley NR, Parrish PR, Lowe JI (1975) Arch. Environ. Contam. Toxicol. *3*:22

Nishahara Y, Utsumi K (1985) Arch. Environ. Contam. Toxicol. *14*:65

O'Connors HB Jr, Wurster CF, Powers CD, Biggs DC, Rowland RG (1978) Science *201*:737

Ofstad EB, Martinsen K (1983) Ambio. *12*:262

Olsson M, Jensen S, Reutergard R (1978) Ambio. *7*:66

Olsson M, Kihlstrom JE, Jensen S, Orberg J (1979) Ambio. *8*:25

Pappageorge W (1983) Haley vs. Monsanto et al., Huron County Michigan, File No. 77-002593, *37*:127

Parkinson A, Robertson LW, Safe S (1980) Biochem. Biophys. Res. Commun. *96*:882

Parkinson A, Safe S (1987) Environ. Toxin. Ser. *1*:49

PavLou SP, Hom W (1979) Ann. N.Y. Acad. Sci. *320*:651

Peakall DB, Lincer JL, Bloom SE (1972) Environ. Hlth. Persp. *1*:103

Platonow NS, Karstad LH (1973) Can. J. Comp. Med. *37*:391

Platonow NS, Liptrap RM, Geissinger HD (1972) Bull. Environ. Contam. Toxicol. *7*:358

Poland A, Greenlee WF, Kende AS (1979) Ann. N.Y. Acad. Sci. *320*:214

Pomerantz I, Burke J, Firestone D, McKinney J, Rooch J, Trotter W (1978) Environ. Hlth. Persp. *24*:133

Reijnders PJH (1980) Neth. J. Sea Res. *14*:30

Rifkind AB, Sassa S, Reyes J, Muschick H (1985) Toxicol. Appl. Pharmacol. *78*:268

Roesijadi GS, Petrocelli R, Anderson JW, Giam CS, Neff GE (1976) Bull. Environ. Contam. and Toxicol. *15*:297

Rogers-Back AM, Clark JJ (1986) Toxicologist. *6*:39

Rosato P, Ferguson DE (1968) Biosci. *18*:783

Rozman K, Pereira D, Iatropoulos MJ (1986) Toxicol. Appl. Pharmacol. *82*:551

Ruzo LO, Zabik MJ, Schuetz RD (1974) J. Am. Chem. Soc. *96*:3809

Safe S (1984) CRC Crit. Rev. Toxicol. *12*:319

Safe S, Bandiera S, Sawyer T, Robertson L, Safe L, Parkinson A, Thomas PE, Ryan DE, Reik LM, Levin WL, Denomme MA, Fujita T (1985a) Environ. Hlth. Persp. *60*:47

Safe S, Safe L, Mullin M (1985b) J. Agric. Fd. Chem. *33*:24

Schmidt T, Risebrough TRW, Gress F (1971) Bull. Environ. Contam. and Toxicol. *6*:235

Schmoldt A, Herzberg W, Benthe HF (1977) Chem.-Biol. Interact. *16*:191

Seegal RF, Bush B, Brosch KO (1985) Neurotox. *6*:13

Shackelford ME, Khan MAQ (1981) Comp. Biochem. Physiol. *70C*:77

Shelton DW, Hendricks JD, Coulombe RA, Bailey GS (1984) J. Toxicol. Environ. Hlth. *13*:649

Shull LR, Bleavins MR, Olson BA, Aulerich RJ (1982) Arch. Environ. Contam. Toxicol. *11*:313

Sipes IG (1987) Environ. Toxin. Ser. *1*:97

Sissons D, Welti D (1971) J. Chromatogr. *60*:15

Skrentny RF, Hemken RW, Dorough HW (1971) Bull. Environ. Contam. Toxicol. *6*:409

Söndergren A (1984) Ambio. *13*:206

Sparling J, Safe S (1980) Toxicol. Lett. *7*:23

Stafford DP, Plapp FW Jr, Fleet RR (1976) Arch. Environ. Contam. Toxicol. *5*:15

Stalling D, Mayer FL (1972) Environ. Hlth. Persp. *1*:165

Standnicki SS, Allen JR (1979) Bull. Environ. Contam. Toxicol. *23*:788

Stegeman JJ (1978) J. Fish. Res. Bd. Canada *35*:668

Stegeman JJ, Kloepper-Sams PJ, Farrington JW (1986) Science *231*:1287

Sugiura K, Kitamura M, Matsumoto E, Goto M (1986) Arch. Environ. Contam. Toxicol. *15*:69

Sundlof SF (1976) MS Thesis. Univ. of Illinois at Urbana-Champaign

Sundström G, Hutzinger O, Safe S (1976) Chemosphere. *5*:267

Suzuki M, Alzawa M, Okand G, Takahashi T (1977) Arch. Environ. Contam. and Toxicol. *5*:343

Swain WR (1983) In: D'ltri and Kamrin, p. 11

Tatsukawa R, Tanabe S (1984) In: Barros et al., p. 99

Torok P (1976) Bull. Environ. Contam. Toxicol. *16*:33

Tryphonas L, Truelove J, Zawidzka Z, Wong J, Mes J, Charbonneau S, Grant DL, Campbell JS (1984) Toxicol. Pathol. *12*:10

Tsushimoto G, Asano S, Trosko JE, Chang C-C (1983) In: D'ltri Hri and Hamrin, p. 241

Tuinstra LGMTh (1984) In: Barros et al., p. 39

Tuinstra LGMTh, Traag WA, Keukens HJ (1980) J. Assoc. Off. Anal. Chem. *63*:952

Tulp MThM, Hutzinger O (1978) Chemosphere. *10*:849

Turner JC (1978) Bull. Environ. Contam. Toxicol. *19*:23

VanLuik A (1984) J. Environ. Qual. *13*:415

Voss JG, Koeman JH, van der Maas FL, TenNoever de Brauw MC, deVos RH Fd. Cosmet. Toxicol. *8*:625

Voss JG, Notenboom-Ram E (1972) Toxicol. Appl. Pharmacol. *23*:563

Webb RG, McCall AC (1972) J. Assoc. Off. Anal. Chem. *55*:746

Westöö G (1974) Ambio. *3*:79

Yamamoto H, Yoshimura H (1973) Chem. Pharm. Bull. (Tokyo) *21*:2237

Zell M, Ballschmiter K (1980) Fres. Z. Analyt. Chem. *304*:337

Zullei N, Benecke G (1978) Bull. Environ. Contam. Toxicol. *20*:786

Mammalian Biologic and Toxic Effects of PCBs

A. Parkinson [1] and S. Safe [2]

This review outlines the mammalian toxicology of PCBs and several classes of related halogenated aromatic hydrocarbons that are persistent environmental contaminants. PCBs comprise 209 chlorinated derivatives of biphenyl, and the toxicology of the individual PCB isomers and congeners is emphasized. Structure-activity relationships for many of the toxic and biochemical effects of PCBs are described. The qualitative aspects of the toxic and biologic effects of PCBs are highly dependent both on the degree of chlorination of the biphenyl nucleus, and the position of the chlorine atoms (i.e., whether they are *ortho, meta* or *para* to the phenyl-phenyl bridge). For example, certain PCB isomers and congeners resemble phenobarbital in their ability to induce rat liver microsomal cytochrome P-450b, whereas others resemble 3-methylcholanthrene in their ability to induce cytochrome P-450c. Furthermore, several PCB isomers and congeners exhibit properties of both these chemicals, and can induce both forms of cytochrome P-450. The ability of certain PCBs to induce cytochrome P-450c is apparently mediated by a high affinity, low capacity cytosolic receptor protein. The affinity with which individual PCBs bind to this receptor follows the same rank order as the potency with which they induce cytochrome P-450c. Furthermore, the affinity with which PCBs and other halogenated hydrocarbons bind to the cytosolic receptor correlates well with their toxic potency. However, a detailed mechanism of toxicity of PCBs and related compounds has not emerged.

[1] Department of Pharmacology, Toxicology and Therapeutics, Kansas University Medical Center, Kansas City, KS 66103, USA
[2] Texas A&M University, College of Veterinary Medicine, Department of Veterinary Physiology and Pharmacology, College Station, TX 77843-4466, USA

1 Introduction

Interest in the biologic and toxic effects of polychlorinated biphenyls (PCBs) can be attributed in large part to the discovery that these highly stable industrial compounds are extremely pervasive and persistent environmental contaminants. Residues of PCBs have been identified in almost every component of the global ecosystem, including human adipose tissue, blood and milk. In addition, as the biologic and toxic effects of PCBs became known, an interest developed in the potential use of PCBs and related halogenated aromatic hydrocarbons to probe the mechanism of various fundamental biologic processes, particularly the mechanism of induction of liver microsomal cytochrome P-450.

Several detailed accounts of the biologic and toxic effects of PCBs and related halogenated aromatic hydrocarbons have been published recently (1–6). This review will highlight some of the basic principles that apply to the biologic and toxic effects of PCBs and will illustrate their usefulness in probing the mechanism of liver microsomal cytochrome P-450 induction.

2 Basic Principles

2.1 PCBs and Halogenated Aromatic Hydrocarbons

PCBs comprise 209 chlorinated derivatives of biphenyl (7). The basic structure is shown in Figure 1 and the distribution of isomers and congeners is given in Table 1. The qualitative aspects of the biologic and toxic effects of PCBs are highly dependent both on the degree of chlorination of the biphenyl nucleus, and on the position of the chlorine atoms (i.e., whether they are *ortho, meta* or *para* to the phenyl-phenyl bridge) (1–6). This is also true of the potency with which PCBs elicit their biologic and toxic effects.

When an equipotent dose is administered, the most toxic PCBs produce a pathologic syndrome similar to that produced by the most toxic members of other classes of halogenated aromatic hydrocarbons, such as polyhalogenated dibenzo-*p*-dioxins, dibenzofurans, azobenzenes, azoxybenzenes, naphthalenes and, to a limited extent, benzenes (1–6, 8–10). These classes of compounds are structurally related, as illustrated in Figure 2. The commonality of the pathologic syndrome

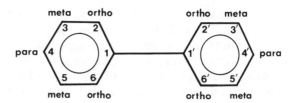

Fig. 1. Structure of biphenyl

Table 1. Distribution of polychlorinated biphenyl isomers and congeners

No. of Cl substituent	1	2	3	4	5	6	7	8	9	10
No. of congeners	3	12	24	42	46	42	24	12	3	1

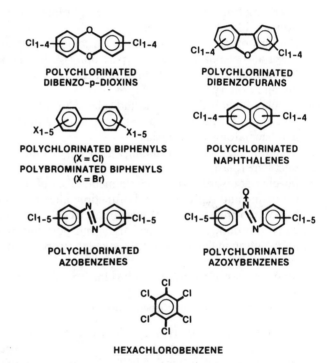

Fig. 2. Structure of some halogenated aromatic hydrocarbons that cause a similar toxic syndrome

produced by the halogenated aromatic hydrocarbons shown in Figure 2 is emphasized for two reasons. First, many of the basic principles that apply to PCBs also apply to the other classes of halogenated aromatic hydrocarbons and, second, some of the effects produced by PCBs were first shown to be caused by other halogenated aromatic hydrocarbons.

In a given species, PCBs and their brominated analogs, PBBs, are less potent than the polyhalogenated dibenzo-p-dioxins, dibenzofurans and the azo- and azoxybenzenes, but are more potent than the polyhalogenated naphthalenes and benzenes. 2,3,7,8-Tetrachlorodibenzo-*p*-dioxin (TCDD) is the most toxic member of all these groups of halogenated aromatic hydrocarbons, and in certain species is one of the most toxic substances known (3, 8–12). TCDD has received considerable attention and serves as a model for understanding the toxicology and mechanism of action of the halogenated aromatic hydrocarbons shown in Figure 2.

2.2 Symptoms of PCB Intoxication

Although there are marked species, age and sex differences (see below), the toxic responses to PCBs and related halogenated aromatic hydrocarbons include:

1. a wasting syndrome: a progressive weight loss which is not simply related to decreased food consumption;
2. skin disorders: acneform eruptions or chloracne, alopecia, edema, hyperkeratosis and blepharitis due to hypertrophy of the Meibomian glands;
3. hyperplasia of the epithelial lining of the extrahepatic bile duct, the gall bladder and urinary tract;
4. lymphoid involution: thymic and splenic atrophy with (a) associated humoral and/or cell-mediated immunosuppression and/or (b) associated bone marrow and haematologic dyscrasias;
5. hepatomegaly and liver damage: necrosis, hemorrhage and intrahepatic bile duct hyperplasia;
6. porphyria: disordered porphyrin metabolism of the cutanea tarda type;
7. endocrine and reproductive dysfunction: altered plasma levels of steroid and thyroid hormones with (a) menstrual irregularities, reduced conception rate, early abortion, excessive menstrual and postconceptional haemorrhage and anovolution in females and (b) testicular atrophy and decreased spermatogenesis in males;
8. teratogenesis: cleft palate and kidney malformations (e.g., renal agenesis); and
9. carcinogenesis: e.g., hepatocarcinoma.

Progressive loss of body weight followed by weakness, debilitation and death characterize the wasting syndrome displayed by most species of animal administered an acutely lethal dose of PCBs. Weight loss is related only in part to decreased food intake and the cause of death is unknown. The chachexia may be responsible for certain reproductive disorders and secondary infections in PCB-exposed animals.

PCBs are weak hepatocarcinogens that may act as promoters rather than initiators of liver cancer (13–19).

2.3 Dose Response

A dose of PCBs administered over time produces a greater toxicity than the same amount of PCBs administered as a single dose. Acute toxicity studies indicate that the dose of commercial mixtures of PCBs or PBBs lethal to 50% of rats ranges from 1–20 g/kg. These high LD_{50} values indicate that PCBs and PBBs are mildly toxic when administered as a single dose (1).

The various symptoms of PCB intoxication do not stem from a single lesion. Consequently, certain lesions, such as thymic atrophy, may be observed at doses of PCBs that have no apparent adverse effect on other organs.

2.4 Time Course

There is a latent period between the time of exposure to PCBs and the onset of certain signs of toxicity. At lethal doses, the latent period between time of exposure and death ranges from about 1 week (e.g., guinea pig) to 1 month (e.g., Rhesus monkey). Interestingly, this latent period cannot be shortened by increasing the dose to a "superlethal" level (1–6).

Some disorders, such as porphyria, become manifest only after several months of exposure to PCBs (20). In contrast, other disorders, such as thymic atrophy, become evident within days of exposure. The extremely long half-life of PCBs means that symptoms of chronic toxicity can develop after exposure has ceased.

2.5 Age and Sex

For a given species, the female is often more susceptible than the male to the toxic (including carcinogenic) effects of halogenated aromatic hydrocarbons and susceptibility to toxicity usually decreases with age (1–6). For example, the LD_{50} for TCDD in adult male rats (60 µg/kg) is more than twice that in either adult female rats (25 µg/kg) or in weanling male rats (25 µg/kg) (8, 12).

Decreased fertility in animals chronically exposed to sublethal levels of halogenated aromatic hydrocarbons are primarily attributable to the female of the species. This may be a consequence of females being more sensitive than males to the toxic effects of PCBs etc.

2.6 Species and Strain Differences

There are marked differences in the sensitivity of various species of animals to the toxic effects of PCBs and related halogenated aromatic hydrocarbons (1–12). Differences in the response of various species to TCDD have been especially well documented (12). The dosage of TCDD lethal to 50% of exposed animals varies more than 3 orders of magnitude, with guinea pigs ($LD_{50} \sim 2$ µg/kg) and hamster LD_{50} 3000–5000 µg/kg) occupying the extremes. Like the guinea pig, various species of bird and mink (21) are very sensitive to TCDD intoxication whereas fish and amphibians are relatively resistant, like the hamster. Rats, rabbits, mice and monkey occupy an intermediate position in their susceptibility to TCDD intoxication as shown in Table 2. A similar rank order to that shown in Table 2 applies to the sensitivity of various species to other toxic effects of TCCD, such as thymic

Table 2. Species differences in the single dose LD_{50} values for TCDD

Species	Route of administration	LD_{50} (μg/kg)
Guinea pig	o.p.	2
Monkey	o.p.	50
Rat (adult male)	i.p.	60
Rabbit	o.p.	115
Mouse C57BL/6J	i.p.	132
DBA/2J	i.p.	620
Hamster	o.p.	5051

Adapted from Ref. 8 and 12

atrophy (1–12). The relative sensitivity of humans to TCDD intoxication is not known.

Various strains of mice differ in their susceptibility to the toxic effects of halogenated aromatic hydrocarbons. The single i.p. dose of TCDD lethal to 50% DBA/2J mice (620 μg/kg) is almost 5 times that for C57BL/6J mice (132 μg/kg), with hybrid (B2D2F1/J) mice displaying an intermediate sensitivity ($LD_{50} \sim$ 300 μg/kg) (22, 23). Such strain differences in mice have been examined in considerable detail to establish genetic factors that influence the biologic and toxic effects of halogenated aromatic hydrocarbons.

Species also differ qualitatively in their response to halogenated aromatic hydrocarbons (1–6, 8, 9). For example, the edematous disorder, hydropericardium, is specific to chickens. Certain qualitative differences seem to reflect differences in the body distribution of halogenated aromatic hydrocarbons. The relatively

Table 3. Comparison of lesions in various species of animals intoxicated with halogenated aromatic hydrocarbons

Species	Lesion Severity[a]					
	Thymus	Liver	Gall bladder	Stomach	Urinary tract	Skin
Guinea pig	+++	±	−	−	+++	−
Mouse	+++	++	−	−	−	++[b]
Rat	+++	++	NA	+	−	−
Hamster	+++	+	++	−	++	+
Chicken	+++	+++	−	+	−	−
Rabbit	+++	+++	−	−	−	+++[c]
Monkey	+++	+	+++	+++	++	+++
Cattle	++	+	++	−	++	+

[a] Severity: ± = minimal; + = mild; ++ = moderate; +++ = marked; NA = not applicable
[b] Present in some strains but not others
[c] Present in ear after local application
Adapted from Ref. 8

high levels of TCDD in the skin of the Rhesus monkey may account for the high sensitivity of this primate to TCDD's acneogenic effects. However, the observation that the guinea pig is resistant to the hapatotoxic effects of TCDD, despite the fact that most of the administered TCDD resides in the liver, indicates that factors beside pharmacokinetic differences are important in determining the qualitative responses to halogenated aromatic hydrocarbons. A summary of severity of various symptoms of halogenated aromatic hydrocarbon intoxication in different mammalian species is given in Table 3. The most consistent symptom of halogenated aromatic hydrocarbon intoxication in all species is thymic atrophy, which results primarily from a loss of cortical lymphocytes (1–12). Thymic atrophy is also one of the most sensitive responses to halogenated aromatic hydrocarbon exposure.

2.7 Role of Metabolic Activation

The toxicity of many chemicals is dependent on their biotransformation to reactive metabolites by cytochrome P-450-dependent monooxygenases or other drug-metabolizing enzymes (24–26). Most evidence indicates that the toxicity of PCBs and related halogenated aromatic hydrocarbons is not dependent on their metabolic activation (27). Indeed the toxicity of PCBs is inversely related to drug-metabolizing activity. For example, females are more susceptible to PCB toxicity than males and the young more susceptible than the old while drug-metabolizing activity is higher in the male and increases with age. Furthermore, induction of drug-metabolizing enzymes by phenobarbital, 3-methylcholanthrene and even TCDD itself, affords some protection against TCDD toxicity. This suggests that the parent halogenated aromatic hydrocarbon, rather than one or more of its metabolites, is the ultimate toxin. Indeed the two major mammalian metabolites of TCDD, namely 2-hydroxy-3,7,8-trichlorodibenzo-p-dioxin and 2-hydroxy-1,3,7,8-tetrachlorodibenzo-p-dioxin, were shown recently to possess less than 0.1% of the toxic potency of TCDD (28).

2.8 In vitro Toxicity

In general, cultured cells and established cell lines are remarkably resistant to the toxic effects of halogenated aromatic hydrocarbons (29). To account for this resistance, it has been proposed that derangement of an integrated metabolic process, possibly one under hormonal regulation, is the cause of the halogenated aromatic hydrocarbon-induced wasting syndrome and lethality (30–36). Although there is no *in vitro* model to probe the mechanism of the wasting syndrome, there are *in vitro* systems to study the porphyrinogenic effects of halogenated aromatic hydrocarbons (based on their ability to inhibit uroporphyrinogen decarboxylase and/or to induce σ-aminolevulinic acid synthetase) (37), to study the dermal toxicity of these compounds (based on their ability to induce keratinization of epidermal cells) (38, 39), and to study their immunotoxic effects (40, 42).

3 Toxicity of Individual PCBs

An evaluation of the toxic effects of individual PCB isomers and congeners has been limited by three problems. First, individual PCBs are not readily available in large quantity, but must be synthesized and isolated by procedures that preclude co-purification of highly active polychlorinated dibenzofurans and di-benzo-p-dioxins. Two, the toxicity of PCBs cannot be evaluated adequately with an *in vitro* system (see above), but must be assessed *in vivo* (which requires much more material). Three, PCBs are relatively non-toxic when administered as a single dose, for which reason large dosages must be given acutely or smaller dosages must be given chronically. Given these problems, it is not surprising that the toxicity of only a few of the 209 PCB isomers and congeners has been evaluated.

Even at the outset, it was suspected that the individual PCB isomers and congeners would differ greatly in their toxic potency. Part of this suspicion was based on the observation that there was considerable variation in toxic potency among the various formulations of industrial PCB mixtures (43, 44). For the most part, these differences persisted even after any contaminating polychlorinated dibenzofurans/dibenzo-p-dioxins were removed from the PCB mixtures by Florisil/alumina chromatography. The finding that the structure of the individual PCBs markedly influence their toxic potency (45–54) was also suspected by analogy with the polychlorinated dibenzo-p-dioxins. It was known that the number and position of the chlorine substituents on the dibenzo-p-dioxin nucleus had a profound influence on toxic potency, with 2,3,7,8-TCDD being the most active member of a series of isomers and congeners whose toxic potency varied over more than 4 orders of magnitude (3, 33). Second, the toxicity of the individual polychlorinated dibenzo-p-dioxins correlated with their ability to induce aryl hydrocarbon hydroxylase (AHH), which measures cytochrome P-450c induction with benzo[a]pyrene as substrate[3]. This same correlation between toxicity and AHH induction was evident among individual polychlorinated dibenzofurans, azobenzenes and azoxybenzenes. In each case, the most active isomers and congeners were approximate isostereomers of TCDD, having two laterally chlorinated aromatic rings in a coplanar configuration with overall dimensions of approximately 3×10 Å.

The toxicity of several PCB isomers and congeners has been evaluated in rats (55, 56) (Table 4). Thymic atrophy was chosen as an end point of toxicity because of its sensitivity, early onset and common occurrence in most species (see above). Body weight loss was also examined as a sign of intoxication. However, the time courses of thymic atrophy and body weight loss are not the same in PCB intoxicated rats, such that only the former is complete with 5 days.

The PCBs listed in Table 4 are divided into two groups, the first of which comprises 6 congeners chlorinated in both *para* positions (4 and 4′), in 0, 1, 2, 3 or all 4 *meta* positions (3, 3′, 5 and 5′) but in none of the positions *ortho* to the phenyl-phenyl bridge (2, 2′, 6 and 6′), as shown in Figure 3. These laterally substituted

---

[3] The hydroxylation of benzo[a]pyrene and the O-dealkylation of 7-ethoxyresorufin are frequently used to monitor the induction of P-450c

Table 4. Toxicity of individual PCB isomers and congeners determined by body weight loss and thymic atrophy in rats

Treatment	Body weight (% decrease)	Thymus weight (% decrease)
Corn oil	0[a]	0[b]
Aroclor 1254	0	31%
Coplanar PCBs		
(A) 4,4-dichlorobiphenyl	0	0
(B) 3,4,4′-trichlorobiphenyl	0	0
(C) 3,4,5,4′-tetrachlorobiphenyl	0	30%
(D) 3,4,3′,4′-tetrachlorobiphenyl	0	30%
(E) 3,4,5,3′,4′-pentachlorobiphenyl	32%	67%
(F) 3,4,5,3′,4′,5′-hexachlorobiphenyl	39%	72%
Mono-ortho chlorinated PCBs		
(G) 2,3,4,5,4′-pentachlorobiphenyl (derived from C)	22%	50%
(H) 3,4,5,2′,4′-pentachlorobiphenyl (derived from C)	0	19%
(I) 2,3,4,3′,4′-pentachlorobiphenyl (derived from D)	0	35%
(J) 2,4,5,3′,4′-pentachlorobiphenyl (derived from D)	0	0
(K) 2,3,4,5,3′,4′-hexachlorobiphenyl (derived from E)	0	35%
(L) 2,3,4,3′,4′,5′-hexachlorobiphenyl (derived from E)	0	15%
(M) 2,4,5,3′,4′,5′-hexachlorobiphenyl (derived from E)	0	0
(N) 2,3,4,5,3′,4′,5′-heptachlorobiphenyl (derived from F)	0	0

One-month-old-male rats were administered a single i.p. injection of PCB at dosages of 125 µmol/kg (E and F), 250 µmol/kg (D), 500 µmol/kg (A, B, C, G–N) or 1500 µmol/kg (Aroclor 1254). Values represent statistically significant (<0.01) decreases in body weight or thymus weight 4 days later
[a] Absolute body weight value was 86 ± 3 g ($\bar{x}\pm$S.E., $n=15$)
[b] Absolute thymus weight value was 320 ± 9 mg ($\bar{x}\pm$S.E., $n=15$)
Adapted from Ref. 55

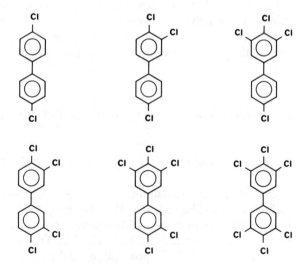

Fig. 3. Structure of the "coplanar" PCBs. These congeners are substituted in both *para* and up to 4 *meta* positions

congeners are referred to as coplanar PCBs because they are devoid of a bulky *ortho*-chloro substituent that would otherwise restrict free rotation about the phenyl-phenyl bond. However, although these PCBs can adopt a coplanar conformation, McKinney and Singh (57) have calculated by the Boltzman distribution law that, at 25 °C, only about 1% of non-*ortho*-substituted biphenyl molecules exist in the coplanar conformation, based on mean of upper (4.5 kcal/mol) and lower (1 kcal/mol) estimates of the energy barrier to rotation.

Three of the so-called coplanar congeners shown in Table 4, namely 3,4,3',4'-tetra-, 3,4,5,3',4'-penta- and 3,4,5,3',4',5'-hexachlorobiphenyl, are the most toxic PCBs known (45, 47, 50, 55, 56, 58–61). The toxic potency of these congeners is highly dependent on the presence of a chloro substituent in both *para* positions. These *para*-chloro substituents, together with the chloro substituents in at least one *meta* position of both phenyl rings, make these particular coplanar PCBs approximate isotereomers of TCDD.

Addition of a single *ortho*-chloro substituent to the coplanar PCBs diminishes their toxic potency, as shown for the second group of PCBs listed in Table 4. Addition of a second *ortho*-chloro substituent diminishes even further the toxic effects of the coplanar PCBs, as determined by body weight and thymus weight loss in rats treated with 500 µmol/kg PCB (55). Recent studies (56) have reported the ED_{50} values for body weight loss and thymic atrophy for several coplanar PCBs and their mono-*ortho* substituted derivatives. There was an excellent correlation between the quantitative and qualitative structure-activity relationships and it was evident that coplanar PCB congeners were the most toxic members of this group of compounds. Moreover, the ED_{50} values for body weight loss (3.3 µmol/kg) and thymic atrophy (~ 1 µmol/kg) for 3,4,5,3',4'-pentachlorobiphenyl in rats demonstrated that TCDD, the most toxic halogenated aromatic hydrocarbon, was only 10–50 times more potent than this coplanar PCB.

Measurements of a variety of end points of toxicity have confirmed that the toxic potency of PCBs is positively correlated with the presence of *para* and *meta*-chloro substituents and inversely correlated with the presence of *ortho*-chloro substituents. An important exception to this generalization is the effect of structure on the tumor-promoting/carcinogenic effects of PCBs (17).

4 The Liver Microsomal Cytochrome P-450 System

4.1 Induction and Multiplicity of Cytochrome P-450 Isozymes

Liver endoplasmic reticulum (microsomes) contains a family of at least 12 cytochrome P-450 isozymes (63–66). The function of these hemoproteins is to catalyze the biotransformation of lipophilic xenobiotics to metabolites that are more readily eliminated from the body (24–26). The cytochrome P-450 system is capable of hydroxylating, epoxidating, dealkylating or oxygenating innumerable xenobiotics, such as drugs, pesticides, environmental pollutants and chemical carcinogens (63–67). Dehydrohalogenation and reduction of certain xenobiotics can also be catalyzed by cytochrome P-450 (67). In addition to xenobiotics, several lipophilic endogenous substrates, such as fatty acids, eicosanoids, sterols, steroid hormones and fat-soluble vitamins, can be hydroxylated by cytochrome P-450 (68, 69). Biotransformation of xenobiotics by cytochrome P-450 is not always a beneficial process; there are many known cases where the metabolites are more toxic or biologically active than the parent compound (24–26, 67). In many cases, the carcinogenic effect of certain xenobiotics is dependent on their conversion by cytochrome P-450 to a reactive, carcinogenic metabolite (26).

A nomenclature for the various forms of cytochrome P-450 has not been established. The broad and overlapping substrate specificity of the individual forms of cytochrome P-450 complicate a traditional nomenclature of these enzymes based on the reactions they catalyze. Ryan et al. (65) have designated the individual forms of cytochrome P-450 in a non-descriptive manner, based on their order of purification. In addition to the 10 forms of rat liver microsomal cytochrome P-450 purified by Ryan et al. (65, 70, 71), and designated cytochromes P-450a – P-450j, the major form of cytochrome P-450 inducible by pregnenolone-16α-carbonitrile (PCN), designated cytochrome P-450p, has been purified and characterized (72–74). A clofibrate-inducible form of cytochrome P-450 (75) and a form of cytochrome P-450 that effectively catalyzes debrisoquine hydroxylation (76) are forms of rat liver microsomal cytochrome P-450 likely distinct from cytochromes P-450a – P-450j and P-450p. The exact number of forms of rat liver microsomal cytochrome P-450 remains unknown (77–80).

The broad and overlapping substrate specificity of the individual forms of cytochrome P-450 complicate not only their nomenclature but also their analysis in a mixture of isozymes, such as that found in liver microsomes. Likewise the spectral, ligand-binding and electrophoretic properties of the individual forms of cytochrome P-450 are too similar to allow each form of cytochrome P-450 to be assayed specifically and unambiguously. It is apparent from Table 5 that a combination of indirect assays can provide considerable information on the profile of cytochrome P-450 isozymes in liver microsomes. However, this information is currently best provided by immunochemical analysis; in which monospecific antibodies provide the specificity necessary to quantitate the levels of an individual form of cytochrome P-450 (77–80).

Several of the cytochrome P-450 isozymes are highly inducible by different xenobiotics as shown in Table 6 for the various rat liver microsomal cytochrome P-450 isozymes. Early studies revealed that a wide variety of structurally diverse xenobiotics that induce rat liver microsomal cytochrome P-450 can be divided into two primary classes (24); one of which is typified by phenobarbital and preferentially induces cytochrome P-450b, and the other of which is typified by 3-methylcholanthrene and preferentially induces cytochrome P-450c (77–79). To a lesser extent, phenobarbital-type inducers also induce cytochromes P-450e (77–79)[4] and P-450p (80), 3-methylcholanthrene-type inducers also induce cy-

---

[4] Evidence suggests that cytochromes P-450b and P-450e are coordinately regulated but that cytochrome P-450b predominates over cytochrome P-450e in Long-Evans rats. However, the precise molar ratio of these two isozymes in liver microsomes from rats treated with different xenobiotics is unknown

Table 5. Effects of phenobarbital or 3-methylcholanthrene treatment of rats on the liver microsomal cytochrome P-450 system

Characteristic	Phenobarbital	3-Methylcholanthrene
CO-ferrocytochrome P-450 complex, λ^{max}	450 nm ($\uparrow$2–3 fold)	448 nm ($\uparrow \sim$ 2 fold)
Ethylisocyanide-ferrocytochrome P-450 complex, 430:455 nm peak ratio	2:1	1:2
Polypeptides intensified on SDS-PAGE	52,00 (P-450b) 52,500 (P-450e) 49,000 (EH) 48,000 (P-450a)	56,000 (P-450c) 52,000 (P-450d) 48,000 (P-450a)
Ethoxycoumarin 0-dealkylation	$\uparrow$	$\uparrow$
Aminopyrine N-demethylation	$\uparrow$	$\leftrightarrow$
Benzphetamine N-demethylation	$\uparrow$	$\leftrightarrow$
Ethylmorphine N-demethylation	$\uparrow$	$\leftrightarrow$
Hexobarbital 3-hydroxylation	$\uparrow$	$\leftrightarrow$
Aldrin epoxidation	$\uparrow\uparrow$	$\leftrightarrow$
Dichloronitroanisole 0-demethylation	$\uparrow\uparrow\uparrow$	$\leftrightarrow$
Pentoxyresorufin 0-dealkylation	$\uparrow\uparrow\uparrow$	$\leftrightarrow$
Testosterone 16β-hydroxylation	$\uparrow\uparrow\uparrow$	$\leftrightarrow$
Benzo[a]pyrene hydroxylation	$\leftrightarrow$	$\uparrow\uparrow$
4-Chlorobiphenyl hydroxylation	$\leftrightarrow$	$\uparrow\uparrow$
Zoxazolamine 6-hydroxylation	$\leftrightarrow$	$\uparrow\uparrow$
Ethoxyresorufin 0-dealkylation	$\leftrightarrow$	$\uparrow\uparrow\uparrow$
NADPH-cytochrome c reductase	$\uparrow$	$\leftrightarrow$
Inhibitors	SKF 525A metyrapone	7,8-benzoflavone ellipticine

$\leftrightarrow$ No increase
$\uparrow$ Increased up to 5 fold
$\uparrow\uparrow$ Increased up to 25 fold
$\uparrow\uparrow\uparrow$ Increased more than 25 fold

Table 6. Inducibility of some rat liver microsomal cytochrome P-450 isozymes

Inducer	Cytochrome P-450 isozyme	Inducibility
Phenobarbital	P-450b	> 50 fold
	P-450e	> 50 fold
	P-450p	up to 20 fold
	P-450a	2–3 fold
3-Methylcholanthrene	P-450c	> 50 fold
	P-450d	up to 20 fold
	P-450a	2–3 fold
Isosafrole	P-450d	$\sim$ 20 fold
	P-450c	$\sim$ 10 fold
	P-450b	up to 20 fold
	P-450a	2–3 fold
Pregnenolone-16α-carbonitrile	P-450p	up to 50 fold

tochrome P-450d (78, 79) and certain xenobiotics from both classes induce cytochrome P-450a (77–79). Exceptions to this classification of cytochrome P-450 inducers are known, including isosafrole (78, 79) and pregnenolone-16α-carbonitrile (80). The major isosafrole-inducible form of cytochrome P-450 is cytochrome P-450d (78, 79) and the major pregnenolone-16α-carbonitrile-inducible form of cytochrome P-450 is cytochrome P-450p (80) (Table VI).

4.2 Mechanisms of Cytochrome P-450 Induction

The induction of cytochrome P-450b by phenobarbital and of cytochrome P-450c by 3-methylcholanthrene is regulated at the transcriptional level (81–87). An increased rate of synthesis largely accounts for the 50-100-fold increase in the concentration of these isozymes after treatment of rats with the appropriate inducer. A high affinity, low capacity cytosolic protein that binds 3-methylcholanthrene-type inducers has been implicated in the induction of cytochrome P-450c (88–91). This receptor is defective in certain strains of mice, such as DBA/2J, which consequently are non-responsive to the inductive effects of 3-methylcholanthrene and other aromatic hydrocarbons (*Ah* non-responsive) (88–92). This contrasts with *Ah* responsive mice, such as C57BL/2J, which contain a normal *Ah* receptor and consequently are responsive to the inductive effects of various polycyclic aromatic hydrocarbons. High doses of TCDD (about 10x that required to induce C57BL/2J mice) can overcome the *Ah* receptor defect and elicit the induction of cytochrome P_1-450 (the mouse counterpart of rat cytochrome P-450c) in *Ah* non-responsive DBA/2J mice. In contrast to TCDD, high doses of 3-methylcholanthrene do not overcome the *Ah* receptor defect in DBA/2J mice; although curiously 3-methylcholanthrene binds to the *Ah* receptor *in vitro* with almost the same affinity as TCDD.

The equivalent of an *Ah* receptor for 3-methylcholanthrene and TCDD has not been found for phenobarbital and other inducers of cytochrome P-450b (93). The cellular mechanism for recognizing phenobarbital-type inducers and for signalling transcription of the cytochrome P-450b and P-450e genes are unknown.

4.3 Induction of Cytochrome P-450 by Industrial PCB Mixtures

A large number of industrial mixtures of PCBs (Aroclor, Phenoclor, Kanechlor and Clophen) and PBBs (FireMaster) have been shown to induce liver microsomal cytochrome P-450 in a variety of species (rats, mice, rabbits, guinea pigs, hamsters, ferret, mink, monkey, fish and birds) (1). There is also evidence for cytochrome P-450 induction in humans accidentally exposed to PCBs (1–6, 94).

The industrial mixture studied most intensely as an inducer of liver microsomal cytochrome P-450 is Aroclor 1254. The effects of Aroclor 1254 treatment on the liver microsomal cytochrome P-450 system in rats are different from those produced by treatment with either phenobarbital or 3-methylcholanthrene. Initially, it was thought that Aroclor 1254 induced a unique form of cytochrome P-450 and, hence, represented a new class of inducer (95). However, it is now known

Table 7. Immunochemical quantitation of cytochrome P-450 isozymes and epoxide hydrolase in liver microsomes from rats treated with phenobarbital, 3-methylcholanthrene or the industrial PCB mixture, Aroclor 1254

Rat treatment (Dose)	nmol Cytochrome P-450/mg microsomal protein					
	Total	P-450a	P-450b + P-450e	P-450c	P-450d	Unknown
Corn oil	0.92	0.08 (8.5%)	0.03 (3.0%)	0.03 (3.0%)	0.06 (6.6%)	0.73 (79%)
Phenobarbital (PB) (4 × 400 μmol/kg)	2.7	0.20 (7.4%)	1.8 (67%)	0.03 (1%)	0.07 (2.6%)	0.60 (22%)
3-Methylcholanthrene (MC) (4 × 100 μmol/kg)	1.9	0.28 (15%)	0.03 (1.6%)	1.1 (58%)	0.80 (42%)	<0
PB + MC (co-administered)	3.4	0.36 (11%)	1.4 (41%)	1.0 (29%)	0.64 (19%)	0
Aroclor 1254[a] (1 × 1500 μmol/kg)	4.2	0.36 (8.7)	1.8 (43)	1.0 (24)	0.83 (20)	0.17 (4)

[a] Based on an average molecular weight of 326

Rats were killed 4 days after a single i.p. injection of Aroclor 1254 or 24 h after 4 consecutive daily i.p. injections of phenobarbital and/or 3-methyl-cholanthrene. Total cytochrome P-450 was determined from the CO-difference spectrum of dithionite-reduced microsomes and immunochemical quantitation of cytochrome P-450 isozymes in liver microsomes was performed as described (55). Numbers in parentheses give the percent total cytochrome P-450 represented by each cytochrome P-450 isozyme. The percentage of unknown cytochrome P-450 represents the arithmetic difference between 100% and the sum of the percentages of each cytochrome P-450 isozyme that the atypical properties of Aroclor 1254 as an inducer of rat liver microsomal cytochrome P-450 stem from its ability to exhibit properties of both phenobarbital- and 3-methylcholanthrene-type xenobiotics and, thus, to induce cytochromes P-450a – P-450e (77, 96–98). This phenobarbital + 3-methylcholanthrene or "mixed" type induction by Aroclor 1254 is shown in Table 7.

Aroclor 1254 is a complex mixture of PCBs and, in 1977–1978, three independent research groups categorized individual PCB isomers and congeners on a structural basis into phenobarbital- or 3-methylcholanthrene-type inducers of liver microsomal cytochrome P-450 (59–61). Unfortunately, the results did not explain the enzyme-inducing properties of Aroclor 1254, which contains only trace to non-detectable levels of the three PCB congeners identified as 3-methylcholanthrenetype inducers, namely 3,4,3',4'-tetra-, 3,4,5,3',4'-penta-, and 3,4,5,3',4',5'-hexachlorobiphenyl. This apparent anomaly is not confined to Aroclor 1254, but is also exhibited by the commercial PBB mixture, fireMaster BP-6 (1, 2). Attempts to identify those components in Aroclor 1254 and fire Master BP-6 that are responsible for the 3-methylcholanthrene-type inducing properties of these halogenated biphenyl mixtures led to the identification of several individual PCB and PBB isomers and congeners that simultaneously exhibit both phenobarbital- and 3-methylcholanthrene-type inducing characteristics (55, 99–110).

These studies with individual PCBs and PBBs have exposed a property that to date is unique to halogenated biphenyls, namely that within this single class of compounds, qualitatively different types of cytochrome P-450 inducers can be constructed by appropriately halogenating the parent hydrocarbon. This property has been very useful to the study of xenobiotic-metabolizing enzymes.

It is important to understand that there was an ulterior motive behind the various attempts to identify those PCB isomers and congeners that were responsible for the 3-methylcholanthrene-type characteristics of Aroclor 1254. Studies with polychlorinated dibenzo-p-dioxin and the initial studies on the structure-activity relationship for PCB isomers and congeners revealed a relationship between toxicity and the ability of individual halogenated aromatic hydrocarbons to exhibit 3-methylcholanthrene-type characteristics (1–6, 33, 34). It was anticipated, therefore, that the development of structure-activity rules for PCBs as inducers of cytochrome P-450 would expedite the search for the toxic PCB isomers and congeners that contaminate the environment.

5 Structure–Activity Relationships

5.1 Coplanar Congeners

Figure 3 shows the structure of 6 coplanar PCB congeners, each of which is chlorinated in both *para* positions and in up to 4 *meta* positions. With the exception of 4,4'-dichlorobiphenyl, which is a weak phenobarbital-type inducer, the coplanar PCBs induce rat liver microsomal cytochromes P-450a, P-450c and P-450d (Table 8). The three most toxic PCB congeners, namely 3,4,3',4'-tetra-, 3,4,5,3',4'-

Table 8. Immunochemical quantitation of cytochrome P-450 isozymes and epoxide hydrolase in liver microsomes from rats treated with polychlorinated biphenyls: Category one, co-planar PCBs

Rat treatment (Dose, μmol/kg)	nmol Cytochrome P-450/mg microsomal protein					
	Total	P-450a	P-450b + P-450c	P-450c	P-450d	Unknown
Corn oil (5 mg/kg)	0.92	0.08 (8.5)	0.03 (3.0)	0.03 (3.0)	0.06 (6.6)	0.73 (79)
3,4,4'-Trichlorobiphenyl (500)	2.51	0.13 (5.0)	0.19 (7.5)	0.68 (27)	0.33 (13)	1.18 (47)
3,4,5,4'-Tetrachlorobiphenyl (500)	2.55	0.28 (11)	0.41 (16)	0.87 (34)	0.48 (19)	0.51 (20)
3,4,3',4'-Tetrachlorobiphenyl (250)	2.36	0.24 (10)	<0.03 (<1.0)	1.0 (44)	0.61 (26)	0.45 (19)
3,4,5,3',4'-Pentachlorobiphenyl (125)	3.85	0.54 (14)	<0.04 (<1.0)	1.2 (30)	1.8 (47)	0.31 (8)
3,4,5,3',4',5'-Hexachlorobiphenyl (125)	4.43	0.62 (14)	<0.05 (<1.0)	1.3 (30)	2.4 (55)	0 (0)

Numbers in parentheses give the percent total cytochrome P-450 represented by each cytochrome P-450 isozyme. The percentage of unknown cytochrome P-450 represents the arithmetic difference between 100% and the sum of the percentages of each cytochrome P-450 isozyme. Adapted from Ref. 55

penta- and 3,4,5, 3',4',5'-hexachlorobiphenyl (see Table 4, exhibit no phenobarbital-type inducing properties, whereas 3,4,4'-tri- and 3,4,5,4'-tetrachlorobiphenyl cause a 6- to 14-fold induction of cytochromes P-450b + P-450e, as shown in Table 8. Based on this qualitative difference, the coplanar PCBs can be divided into two groups: with 3,4,3',4'-tetra-, 3,4,5,3',4'-penta- and 3,4,5,3',4',5'-hexachlorobiphenyl in group I and with 4,4'-di-, 3,4,4'-tri- and 3,4,5,4'-tetrachlorobiphenyl in group II. A similar subdivision of the coplanar PCBs can be made on the basis of toxicity (with group I being more toxic than group II as shown in Table 4); on the basis of potency (with group I being at least 100 times more effective than group II as inducers of cytochrome P-450c-dependent monooxygenase activity in cultured rat hepatoma cells), and on the basis of receptor binding affinity (with group I being at least 70 times more effective than group II in competitively displacing TCDD from a high affinity rat liver cytosol receptor protein) (55, 56, 111, 112).

Group I coplanar PCBs are approximate isostereomers of TCDD, which explains in part why their biologic/toxic potency is greater than that of the Group II coplanar PCBs. However, the Group I coplanar PCBs are less potent than TCDD, which is attributable in large part to the free rotation of the two phenyl rings in PCBs. Indeed, when 3,4,3',4'-tetrachlorobiphenyl is locked into the corresponding planar structure, 2,3,6,7-tetrachlorobiphenylene, the ED_{50} for cytochrome P-450c induction and K_D for receptor binding approach very closely the corresponding TCDD values (60).

5.2 *Mono-ortho* Chlorinated Derivatives of Coplanar PCBs

The addition of a single *ortho*-chloro substituent to the higher chlorinated coplanar PCBs (3,4,5,4'-tetra-, 3,4,3',4'-tetra-, 3,4,5,3',4'-penta- and 3,4,5,3',4',5'-hexachlorobiphenyl) gives a series of 8 derivatives, as shown in Figure 4. The toxicity

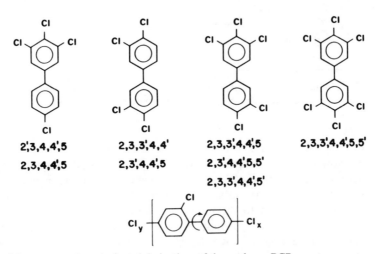

Fig. 4. Structure of the mono-*ortho*-substituted derivatives of the coplanar PCBs

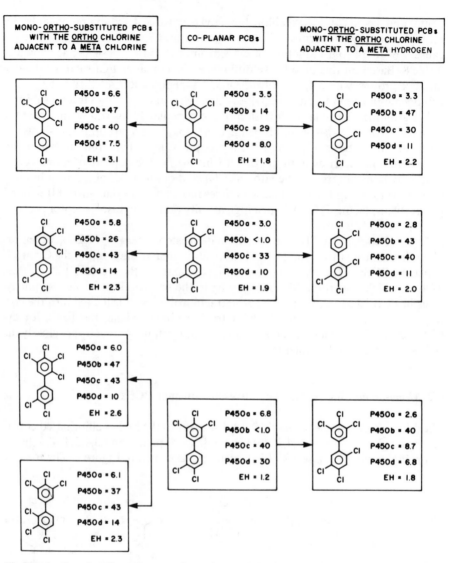

Fig. 5. Induction of rat liver microsomal cytochrome P-450 isozymes and epoxide hydrolase (EH) by coplanar PCBs and their mono-*ortho*-chloro derivatives. From Ref. 55

of the coplanar PCBs is reduced by introduction of an *ortho*-chloro substituent, as shown in Table 4. This loss of toxic potency is accompanied by a qualitative shift in cytochrome P-450 induction. As shown in Figure 5, all of the mono-*ortho*-substituted PCBs are mixed-type inducers (phenobarbital + 3-methylcholanthrene) in contrast to the parent coplanar PCBs, which are predominantly 3-methylcholanthrene-type inducers (see Table 8). Studies on 2,3,4,5,3′,4′,5′-heptachlorobiphenyl have shown that the metabolic removal of the *ortho*-chloro substituent to give the coplanar PCB, 3,4,5,3′,4′,5′-hexachlorobiphenyl, does not occur and, hence, does not account for the 3-methylcholanthrene-type characteristics of this mono-*ortho* chlorinated PCB (112a). The mono-*ortho* chlorinated derivatives of the coplanar PCBs are a highly unusual class of compounds, being potent and effective inducers of both the phenobarbital- and 3-methylcholanthrene-inducible forms of cytochrome P-450.

The introduction of a single *ortho*-chloro substituent into the biphenyl nucleus results in decreased coplanarity between the two phenyl rings due to steric interactions between the bulky *ortho*-chloro and hydrogen substituents. It was initially proposed (59–61) that the reduction in coplanarity would render *ortho*-chlorinated PCBs incapable of interacting with the *Ah* receptor and inducing cytochrome P-450c. The results shown in Figure 5 indicate that the presence of an *ortho*-chloro substituent by no means eliminates the ability of PCBs to induce cytochrome P-450c. Studies by Bandiera et al. (112) have shown the mono-*ortho* chlorinated PCBs in Figure 4 compete effectively with [^{3}H]-TCDD for binding to its receptor in rat liver cytosol.

The mono-*ortho*-chloro substituted PCBs shown in Fig. 4 can be divided into two groups: one has the *ortho*-chloro adjacent to a *meta*-chloro substituent (as in the 2,3,4- and 2,3,4,5-substituted PCBs) whereas the other has the *ortho*-chloro adjacent to a *meta*-hydrogen (as in the 2,4- and 2,4,5-substituted PCBs). In general the former group is more toxic and more potent an inducer of cytochrome P-450c than the latter group (55, 56, 100, 111, 112). This was an unexpected observation because, in the former group of PCBs, the *ortho*-chloro substituent is buttressed by the adjacent *meta*-chloro, forcing it toward the phenyl-phenyl bridge. The ability to adopt a coplanar configuration is impeded more by a buttressed *ortho*-chloro substituent than by a nonbuttressed one. Consequently, if coplanarity of laterally substituted PCBs were the only criterion necessary to induce cytochrome P-450c, it would be predicted that the first group of mono-*ortho*-substituted PCBs (buttressed) would be less active than the second group (non-buttressed). To explain the converse observation, McKinney and colleagues (57, 57a) proposed that the ability of PCBs to compete with TCDD for binding to the cytosolic receptor that mediates cytochrome P-450c induction is dependent on both coplanarity of the two phenyl rings and net polarizability of the biphenyl. Results obtained with various PCBs support this hypothesis, although some exceptions have been identified in experiments with PBBs (55).

5.3 Di-ortho Chlorinated Derivatives of Coplanar PCBs

In the previous section, it was shown that addition of a single *ortho*-chloro substituent to the coplanar PCBs shown in Fig. 3 has two general effects: one is to decrease the toxicity of the coplanar PCBs, as well as their ability to induce cytochrome P-450c, and the other is to increase their ability to induce cytochrome P-450b. This trend toward phenobarbital-type characteristics and away from toxic, 3-methylcholanthrene-type characteristics continues as two or more *ortho*-chloro substituents are added to the coplanar PCBs. This trend is illustrated in Figure 6. Some di-*ortho* substituted derivatives of the coplanar PCBs such as 2,4,5,2′,4′,5′-hexachlorobiphenyl are "pure" phenobarbital-type inducers. In con-

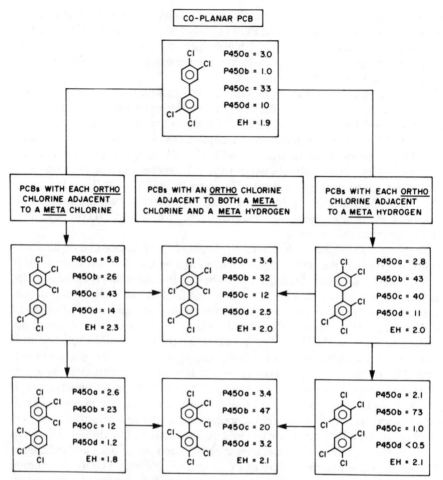

Fig. 6. Induction of rat liver microsomal cytochrome P-450 isozymes and epoxide hydrolase (EH) by mono- and di-*ortho*-chloro derivatives of a coplanar PCB. From Ref. 55

trast, the parent coplanar PCB, 3,4,3',4'-tetrachlorobiphenyl is a "pure" 3-methylcholanthrene-type inducer.

As shown in Figure 7, there are 13 possible di-ortho derivatives of the most active coplanar PCBs. At least 5 of these di-*ortho* derivatives can compete with TCDD for receptor binding sites in rat liver cytosol and induce cytochrome P-450c (they are 2,3,4,2',3',4'-hexa-, 2,3,4,6,3',4'-hexa-, 2,3,4,2',4',5'-hexa-, 2,3,4,5,6,4'-hexa- and 2,3,4,5,2',3',4'-heptachlorobiphenyl). Conflicting reports that other members of the series of the 13 PCBs can induce cytochrome P-450c were resolved in part when Goldstein et al. (113) demonstrated the presence of 2,3,7,8-tetrachlorodibenzofuran as a contaminant in commercially available 2,4,5,2',4',5'-hexachlorobiphenyl. The ability of the 5 above-mentioned di-*ortho*-substituted PCBs to induce rat liver microsomal cytochrome P-450c is not due to the presence of highly active contaminants, which indicates that, in certain cases,

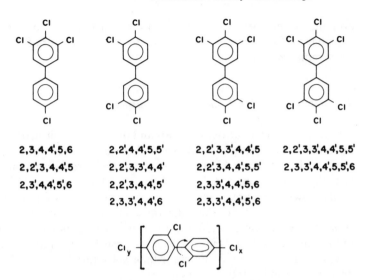

Fig. 7. Structure of the di-*ortho*-substituted derivatives of the coplanar PCBs

Table 9. Summary of structure-function relationships

PCB structures (n)	Cytochrome P-450 induction[a] (% of control)		Relative activity (% of control)		
	P-450c + P-450d	P-450b + P-450e	Aryl hydrocarbon hydroxylase induction		Receptor binding
			In vivo[b]	*In vitro*[c]	
Coplanar PCBs – group I[f]	4100–1800	No induction	+++	100–1	100-35
Coplanar PCBs – group II[g]	1500–1100	1400– 600	++	3×10^{-2} [h]	0.5[h]
Mono-*ortho*-coplanars	2400– 750	4700–2600	++	$0.3–2.4 \times 10^{-5}$	6–1.5
Di-*ortho*-coplanars	900– 250	6300–1000	+	Inactive	<0.3[e]
2,4,5,2′,4′,5′-Hexachlorobi-phenyl	No induction	7300	Inactive	Inactive	<0.3[e]
2,3,7,8-TCDD	3500	No induction	+++++ 400		2500

[a] Male Long-Evans rats
[b] Male Wistar rats
[c] Rat hepatoma H-4-II-E cells
[d] Determined by the competitive displacement of [³H]TCDD bound to liver cytosol from male Wistar rats
[e] Represents nonspecific binding
[f] 3,4,3′,4′-Tetra-, 3,4,5,3′,4′-penta and 3,4,5,3′,5′,5′-hexachlorobiphenyl
[g] 3,4,4′-Tri- and 3,4,5,4′-tetrachlorobiphenyl
[h] Determined only for 3,4,5,4′-tetrachlorbiphenyl
Adapted from Ref. 114

even the presence of two *ortho*-chloro substituents does not abolish the ability of coplanar PCBs to bind to the *Ah* receptor. These structure-function relationships are summarized in Table 9.

5.4 Mixtures of PCB Isomers and Congeners

The systematic evaluation of mono- and di-*ortho* substituted coplanar PCBs was undertaken to explain the biologic and toxic properties of industrial PCB mixtures, such as Aroclor 1254. In contrast to the parent coplanar PCBs, which are virtually absent from Aroclor 1254, many of the mono- and di-*ortho* substituted PCBs shown in Figures 4 and 7 are major components of industrial PCB mixtures (115–117). The ability of these derivatives to induce several forms of rat liver microsomal cytochrome P-450 and to elicit a toxic syndrome qualitatively similar to that caused by TCDD intoxication suggests that the mono- and di-*ortho* derivatives of the coplanar PCBs contribute substantially to the biologic and toxic properties of industrial PCB mixtures.

The importance of structure on the biologic and toxic effects of PCB isomers and congeners has ramifications for evaluating the environmental impact and potential health threat posed by PCBs. The composition of PCB mixtures changes dramatically in the environment, particularly in the food chain (1, 7, 94). It has been shown, for example, that the mixture of PCBs secreted in human milk differs significantly from the PCBs in the industrial mixtures that contaminate the environment. The mixture of PCBs in breast milk is considerably enriched in many of the mono- and di-*ortho* substituted analogs illustrated in Figures 4 and 7. The preferential metabolism and subsequent excretion of lower chlorinated PCBs and the poor absorption of higher chlorinated PCBs likely accounts for this relative enrichment. As a result of this enrichment in the mono- and di-*ortho* substituted PCBs, a mixture of PCBs secreted in human breast milk is 5- to 10-times more biologically (toxicologically?) active than the industrial PCB mixture (118–119). Clearly, a knowledge of isomeric and congeneric content is required to make a meaningful assessment of the potential risk posed by mixtures of PCBs.

The fact that the environment is contaminated with mixtures of PCBs raises the issue of whether or not one PCB isomer or congener can influence the biologic and/or toxic effects of another. The practical concern here is whether the toxicity of an individual PCB can be enhanced through synergistic interaction with one or more other PCBs. Two potential mechanisms of synergistic interaction between individual PCBs have been identified. The first involves a sparing effect, whereby a non-toxic PCB occupies binding sites (non-specific binding sites, receptors, proteins and/or enzymes) and decreases the removal of a toxic PCB via these pathways (120). A second possibility involves an interaction at the level of the cytosolic *Ah* receptor (121, 122). Recent studies have demonstrated that specific PCB congeners elevate hepatic *Ah* receptor levels in the immature male Wistar rat and C57BL/6 mice (122, 123). It has also been shown that the ability of PCBs to elevate hepatic receptor levels was dependent on their structure. The coplanar 3,4,3′,4-tetra- and 3,4,5,3′,4′-pentachlorobiphenyls, which bind with high affinity to the receptor protein, do not elevate hepatic receptor levels, whereas several di-

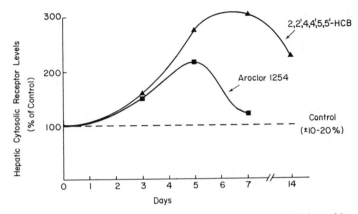

Fig. 8. Elevation of liver cytosolic *Ah* receptor levels in rats treated with 2,4,5,2′,4′,5′-hexachlorobiphenyl or Aroclor 1254

ortho substituted PCBs, which primarily exhibit PB-type induction activity and low binding affinities for the receptor protein (112), significantly elevate hepatic cytosolic *Ah* receptor levels. Mixed-type monooxygenase enzyme inducers (including Aroclor 1254) which bind with moderate affinity to the *Ah* receptor also elevate hepatic receptor levels. The effects of Aroclor 1254 and the PB-type inducer, 2,4,5,2′,4′,5′-hexachlorobiphenyl, on liver cytosolic *Ah* receptor levels in immature male Wistar rats are illustrated in Figure 8.

Previous studies with steroid hormones have demonstrated synergistic interactive effects that can be rationalized in terms of altered hormone receptor levels (124). For example, the increased responsiveness of the chick oviduct to progesterone following estrogen treatment is related, in part, to estrogen-dependent increases and/or changes in the progesterone receptor in the oviduct (124).

Recently we have shown comparable interactive effects with PCB congeners (125). Pretreatment of immature male Wistar rats with 2,4,5,2′,4′,5′-hexachloro-

Table 10. Interactive effects of PCBs on the induction of rat liver microsomal cytochrome P-450c

Treatment (dosage, μmol/kg	Micosomal enzyme activity[a]	
	Benzo[a]pyrene	Ethoxyresorufin O-deethylase
Corn oil	0.088	0.277
2,4,5,2′,4′,5′,-HCB[b] (300)	0.121	0.530
3,4,5,3′,4′-PCB (0.01)	0.486	3.60
3,4,5,3′,4′-PCB+2,4,5,2′,4′,5′-HCB[c]	0.887	6.03
3,4,5,3′,4′,5′-HCB (125)	0.676	6.89
3,4,5,3′,4′,5′-HCB+2,4,5,2′,4′,5′-HCB[c]	1.06	10.5

[a] nmol product/mg protein/min
[b] Hexa- and pentachlorobiphenyl are abbreviated HCB and PCB, respectively
[c] Administered 7 days prior to treatment with 3,4,5,3′,4′-PCB or 3,4,5,3′,4′,5′-HCB

biphenyl (300 µmol/kg) followed by a challenge with a submaximal inducing dose of 3,4,5,3′,4′-penta- or 3,4,5,3′,4′,5′-hexachlorobiphenyl resulted in non-additive induction of hepatic microsomal benzo[a]pyrene hydroxylase and ethoxyresorufin O-deethylase activities (Table 10). Based on PCB structure-activity relationships for receptor elevation and monooxygenase enzyme induction, it is apparent that commercial and environmental mixtures of PCBs contain both classes of compounds. Therefore, it is likely that at some dose levels of these mixtures their hepatic microsomal enzyme induction activities may be dependent on the types of interactive effects exemplified in Table 10. Future studies are in progress to determine if receptor elevation can play a role in the hepatic and extrahepatic toxicity of PCBs and related toxic halogenated aryl hydrocarbons.

6 Acknowledgements

A.P. gratefully acknowledges the financial support from NIH grant ES 03765 and ES 00166 and the PMA, Speas and Flossie West Foundation. S.S. appreciates the financial support from the Texas Agricultural Experiment Station, The Chester Reed endowment, the Natural Sciences and Engineering Research Council, and the National Institutes of Health.

We thank Ms. M. Floyd for her assistance in the preparation of this review, as well as the many colleagues with whom we work and collaborate.

7 References

1. Safe S (1984) CRC Crit. Rev. Toxicol. *13*:319
2. Parkinson A, Safe S (1981) Toxicol. Environ. Chem. Rev. *4*:1
3. Poland A, Knutson IC (1982) Ann. Rev. Pharmacol. Toxicol. *22*:517
4. Kimbrough RD (1974) CRC Crit. Rev. Toxicol. *2*:445
5. Lai DY (1984) J. Environ. Sci. Health *C2*:135
6. Kimbrough RD (ed) (1980) Halogenated Biphenyls, Terphenyls, Naphthalenes, Dibenzo-dioxins and Related Products. Elsevier/North Holland Biomedical Press, Amsterdam
7. Hutzinger O, Safe S, Zitko V (1974) The Chemistry of PCBs. CRC Press, Boca Raton, Florida
8. McConnell EE (1985) Environ. Health Perspect. *60*:29
9. McConnel EE, Moore JA (1979) Ann. N.Y. Acad. Sci. *320*:139
10. Allen JR, Hargraves WA, Hsia MTS, Lin FSD (1979) Pharmacol. Ther. *7*:513
11. Neal RA, Beatty PW, Gasiewicz TA (1979) Ann. N.Y. Acad. Sci. *320*:204
12. Neal RA (1985) Environ. Health Perspect. *60*:41
13. Allen JR, Norback DH (1977) Carcinogenic potential of the polychlorinated biphenyls. In: Hiatt HH, Watson JD, Winsten JA (eds) Origins of Human Cancer. Cold Spring Harbor Laboratory, Cold Spring Harbor, New York, p 173
14. Kimbrough RD, Squire RA, Linder RE, Strandberg JD, Montali RJ, Burse VW (1975) J. Natl. Cancer Inst. *55*:1453
15. Pereira MA, Herren SL, Britt AL, Khoury MM (1982) Cancer Lett. *15*:185

16. Deml E, Oesterle D (1982) Carcinogenesis *3*:1449
17. Sleight S (1985) Environ. Health Perspect. *60*:35
18. Ward JM (1985) Environ. Health Perspect. *60*:89
19. Norback DH, Weltman RH (1985) Environ. Health Perspect. *60*:97
20. Strik JJTWA, Debets FMH, Koss G (1980) In: Kimbrough RD (ed) Halogenated Biphenyls, Terphenyls, Napthalenes, Dibenzodioxins, and Related Products. Elsevier/North-Holland Biomedical Press, Amsterdam, p 19
21. Ringer RK, Auerlich RJ, Bleavins MR (1981) In: Khan MAQ, Stanton RH (eds) Toxicology of Halogenated Hydrocarbons – Health and Ecological Effects. Pergamon Press, N.Y., p 329
22. Gasiewicz TA, Geiger LE, Rucci G, Neal RA (1983) Drug Metab. Dispos. *11*:397
23. Poland A, Glover E (1980) Mol. Pharmacol. *17*:86
24. Conney AH (1967) Pharmacol. Rev. *19*:317
25. Gillette JR, Davis DC, Sasame HA (1972) Ann. Rev. Pharmacol. *12*:57
26. Conney AH (1982) Cancer Res. *42*:4875
27. Beatty PW, Vaughn WK, Neal RA (1978) Toxicol. Appl. Pharmacol. *45*:513
28. Mason G, Safe S (1986) Toxicology. In press
29. Knutson JC, Poland A (1980) Toxicol. Appl. Pharmacol. *54*:377
30. Seefeld MD, Peterson RE (1983) 2,3,7,8-Tetrachlorodibenzo-p-dioxin-induced weight loss: A proposed mechanism. In: Tucker RE, Young AL, Gray AP (eds) Human and Environmental Risks of Chlorinated Dioxins and Related Compounds. Plenum Press, New York, p 405
31. Matsumura F (1983) Pharmac. Ther. *19*:195
32. Rozman K (1984) Biochem. Biophys. Res. Commun. *125*:996
33. Poland A, Greenlee WF, Kende AS (1979) Ann. N.Y. Acad. Sci. *320*:214
34. Poland A, Knutson J, Glover E (1983) A consideration of the mechanism of action of 2,3,7,8-tetrachlorodibenzo-p-dioxin and related halogenated aromatic hydrocarbons. In: Tucker RE, Young AL, Gray AP (eds) Human and Environmental Risks of Chlorinated Dioxins and Related Compounds. Plenum Press, New York, p 539
35. Kociba RJ, Schwetz BA (1982) Drug Metab. Rev. *13*:387
36. Sweeny GD, Jones KG (1983) Studies on the mechanism of action of hepatotoxicity of 2,3,7,8-tetrachlorodibenzo-p-dioxin (TCDD) and related compounds. In: Tucker RE, Young AL, Gray AP (eds) Human and Environmental Risks of Chlorinated Dioxins and Related Compounds. Plenum Press, New York, p 415
37. Sano S, Kawanishi S, Seki Y (1985) Environ. Health Perspect. *59*:137
38. Knutson JC, Poland A (1980) Cell. *22*:27
39. Osborne R, Greenlee WF (1985) Toxicol. Appl. Pharmacol. *77*:434
40. Dencker L, Hassoun E, D'Argy R, Alm G (1985) Mol. Pharmacol. *27*:133
41. Vos JG, Faith RE, Luster MI (1981) Immune alterations. In: Kimbrough RD (ed) Halogenated Biphenyls, Terphenyls, Naphthalene, Dibenzodioxins and Related Products. Elsevier/North Holland, Amsterdam, p 241
42. Faith RE, Luster MI, Vos JG (1980) Effects on immunocompetency by chemicals of environmental concern. In: Hodgson E, Bend JR, Rhilpot RM (eds) Reviews in Biochemical Toxicology 2. Elsevier/North Holland, Amsterdam, p 173
43. Goldstein JA, Hickman P, Burse VW, Bergman H (1975) Toxicol. Appl. Pharmacol. *32*:461
44. Vos JG, Koeman JH (1970) Toxicol. Appl. Pharmacol. *17*:650
45. Goldstein JA (1979) Ann. N.Y. Acad. Sci. *320*:164
46. McKinney JD, Chae K, Gupta BN, Moore JA, Goldstein JA (1976) Toxicol. Appl. Pharmacol. *36*:65
47. Kohli KK, Gupta BN, Albro PW, Mukhtar H, McKinney JD (1979) Chem. Biol. Interact. *25*:139
48. Biocca M, Gupta BN, Chae K, McKinney JD, Moore JA (1981) Toxicol. Appl. Pharmacol. *58*:461
49. Marks TA, Kimmel GA, Staples RE (1981) Toxicol. Appl. Pharmacol. *61*:269
50. Yoshihara S, Kawano K, Yoshimura H, Kuroki H, Masuda Y (1971) Chemosphere. *8*:531

51. Ax RL, Hansen LG (1975) Poult. Sci. *54*:895
52. Hansell MM, Ecobichon DJ (1974) Toxicol. Appl. Pharmacol. *28*:418
53. Aulerich RJ, Bursian SJ, Breslin WI, Olson BA, Ringer RK (1985) J. Toxicol. Environ. Health. *15*:63
54. Yoshimura H, Yoshihara S, Ozawa N, Miki M (1979) Ann. N.Y. Acad. Sci. *320*:179
55. Parkinson A, Safe S, Robertson L, Thomas PE, Ryan DE, Reik LM, Levin W (1983) J. Biol. Chem. *258*:5967
56. Leece B, Denomme MA, Towner R, Li SMA, Safe S (1985) J. Toxicol. Environ. Health. *16*:379
57. McKinney JD, Singh P (1981) Chem. Biol. Interact. *33*:271
57a. Albro PW, McKinney JD (1981) Chem. Biol. Interact. *34*:373
58. Goldstein JA (1980) Structure-activity relationship for the biochemical effects on the relationship to toxicity. In: Kimbrough R (ed) Halogenated Biphenyls, Terphenyls, Naphthalenes, Dibenzodioxins and Related Products. Elsevier/North Holland Biomedical Press, p 151
59. Goldstein JA, Hickamn P, Bergman H, McKinney JD, Walker MP (1977) Chem. Biol. Interact. *17*:69
60. Poland A, Glover E (1977) Mol. Pharmacol. *13*:924
61. Yoshimura H, Ozawa N, Saeki S (1978) Chem. Pharm. Bull. *26*:1215
62. Yoshimura H, Yoshihara S, Koga N, Nagata K, Wada I, Kuroki J, Hokama Y (1985) Environ. Health Perspect. *59*:113
63. Johnson EF (1979) Multiple forms of cytochrome P-450: criteria and significance. In: Hodgson E, Bend JR, Philpot RM (eds) Reviews in Biochemical Toxicology. Vol. 1. Elsevier/North Holland, New York
64. Gungerich FP (1979) Pharmacol. Ther. *6*:99
65. Ryan DE, Thomas PE, Reik LM, Levin W (1982) Xenobiotica *12*:727
66. Guengerich PF, Dannan GA, Wright ST, Martin V, Kaminsky LS (1982) Xenobiotica. *12*:701
67. Guengerich FP, Liebler DC (1985) CRC Crit. Rev. Toxicol. *14*:259
68. Conney AH, Kuntzman R (1971) Handbook Expt. Pharmacol. *28*:401
69. Coon MJ, Koop DR (1983) The Enzymes. *16*:645
70. Ryan DE, Iida S, Wood AW, Thomas PE, Lieber CS, Levin W (1984) J. Biol. Chem. *259*:1259
71. Ryan DE, Ramanathan L, Iida S, Thomas PE, Haniu M, Shively JE, Lieber CS, Levin W (1985) J. Biol. Chem. *260*:6385
72. Elshourbagy NA, Guzelian PS (1980) J. Biol. Chem. *225*:1279
73. Wrighton SA, Maurel P, Schuetz EG, Watkins PB, Young B, Guzelian PS (1985) Biochemistry. *24*:2171
74. Waxman DJ, Dannan GA, Guengerich FP (1985) Biochemistry. *24*:4409
75. Gibson GG, Orton TD, Tamburini PP (1982) Biochem. J. *203*:161
76. Larrey D, Distlerath LM, Dannan GA, Wilkinson GR, Guengerich FP (1984) Biochemistry. *23*:2787
77. Thomas PE, Reik LM, Ryan DE, Levin W (1981) J. Biol. Chem. *256*:1044
78. Thomas PE, Reik LM, Ryan DE, Levin W (1983) J. Biol. Chem. *248*:4590
79. Guengerich FP, Dannan GA, Wright ST, Martin MV, Kaminsky LS (1982) Biochemistry. *21*:6019
80. Heuman DM, Gallagher EJ, Barwick JL, Elshourbagy NA, Guzelian PS (1982) Mol. Pharmacol. *21*:753
81. Colbert RA, Bresnick E, Levin W, Ryan DE, Thomas PE (1979) Biochem. Biophys. Res. Commun. *91*:886
82. Dubois RN, Waterman MR (1979) Biochem. Biophys. Res. Commun. *90*:150
83. Hardwick JP, Gonzalez FJ, Kasper CB (1983) J. Biol. Chem. *258*:8081
84. Adesnik M, Bar-Nun S, Maschio F, Zunich M, Lipman A, Bard E (1981) J. Biol. Chem. *256*:10340
85. Bresnick E, Brosseau M, Levin W, Reik L, Ryan DE, Thomas PE (1981) Proc. Natl. Acad. Sci. USA *78*:4083
86. Lippman-Morville A, Thomas PE, Levin W, Reik L, Ryan DE, Raphael C, Adesnik M (1983) J. Biol. Chem. *258*:3901

87. Kawajiri K, Gotoh O, Tagashira Y, Sogawa K, Fujii-Kuriyama Y (1984) J. Biol. Chem. *259*:10145
88. Poland A, Gover E, Kende AS (1976) J. Biol. Chem. *251*:4936
89. Okey AB, Bondy GP, Mason ME, Kahl GF, Eisen HJ, Guenthner TM, Nebert DW (1979) J. Biol. Chem. *254*:11636
90. Okey AB, Bondy GP, Mason ME, Nebert DW, Forster-Gibson C, Mucan J, Dufresne MJ (1980) J. Biol. Chem. *255*:11415
91. Greenlee WF, Poland A (1979) J. Biol. Chem. *254*:9814
92. Poland A, Glovert E (1974) Mol. Pharmacol. *11*:389
93. Poland A, Mak I, Glover E, Boatman RJ, Ebetino FH, Kende AS (1980) Mol. Pharmacol. *18*:571
94. Higuchi K (ed) (1976) PCB Poisoning and Pollution. Kodansha, Tokyo
95. Alvares AP, Bickers DR, Kappas A (1973) Proc. Natl. Acad. Sci. USA *70*:1321
96. Alvares AP, Kappas A (1977) J. Biol. Chem. *252*:6373
97. Ryan DE, Thomas PE, Levin W (1977) Mol. Pharmacol. *13*:521
98. Ryan DE, Thomas PE, Korzeniowski D, Levin W (1979) J. Biol. Chem. *254*:1365
99. Parkinson A, Cockerline R, Safe S (1980) Biochem. Pharmacol. *29*:259
100. Parkinson A, Cockerline R, Safe S (1980) Chem. Biol. Interact. *29*:277
101. Parkinson A, Robertson L, Safe L, Safe S (1981) Chem. Biol. Interact. *31*:1
102. Parkinson A, Robertson L, Safe L, Safe S (1981) Chem. Biol. Interact. *30*:271
103. Parkinson A, Robertson L, Safe S (1980) Life Sci. *27*:2333
104. Stonard MD, Grieg JB (1976) Chem. Biol. Interact. *15*:365
105. Denomme MA, Bandiera S, Lambert I, Copp L, Safe L, Safe S (1983) Biochem. Pharmacol. *32*:2955
106. Dannan GA, Moore RW, Besaw LC, Aust JD (1978) Biochem. Biophys. Res. Commun. *85*:51
107. Moore RW, Dannan GA, Aust SD (1980) Structure function relationships for the pharmacological and toxicological effects and metabolism of polybrominated biphenyl congeners. In: Bhatnagar RS (ed) Molecular Basis of Environmental Toxicity, Chapt. 8. Ann Arbor Science, Ann Arbor, Michigan
108. Robertson LW, Parkinson A, Campbell MA, Safe S (1982) Chem. Biol. Interact. *42*:53
109. Robertson LW, Parkinson A, Safe S (1980) Biochem. Biophys. Res. Commun. *92*:175
110. Dannan GA, Guengerich FP, Kaminsky LS, Aust SD (1983) J. Biol. Chem. *258*:1282
111. Sawyer T, Safe S (1982) Toxicol. Lett. 87
112. Bandiera S, Safe S, Okey AB (1982) Chem. Biol. Interact. *39*:259
112a. Corbett J, Albro PW, Chae K, Jordan S (1982) Chem. Biol. Interact. *39*:331
113. Goldstein JA, Haas JR, Linko P, Harvan DJ (1978) Drug Metab. Dispos. *6i*:258
114. Safe S, Bandiera S, Sawyer T, Robertson LW, Safe L, Parkinson A, Thomas PE, Ryan DE, Reik LM, Levin W, Denomme MA, Fujita T (1985) Environ. Health Perspect. *60*:47
115. Sissons D, Welti D (1971) J. Chromatogr. *60*:15
116. Ballschmiter K, Zell M (1980) Fresenius Z. Anal. Chem. *302*:20
117. Mullin MD, Pochini CM, Safe SH, Safe LM (1983) Analysis of PCBs using high resolution capillary gas chromatography. In: D'Itri FM, Kamrin MA (eds) PCBs: Human and Environmental Hazards Ann Arbor Science, Ann Arbor, Michigan, p 165
118. Parkinson A, Robertson LW, Safe S (1980) Biochem. Biophys. Res. Commun. *96*:882
119. Gyorkos J, Denomme MA, Leece B, Homonko K, Valli E, Safe S (1985) Can. J. Phys. Pharmacol. *63*:36
120. McKinney JD, Chae K, McConnell EE, Birnbaum LS (1985) Environ. Health Perspect. *60*:57
121. Okey AB, Vella LM (1984) Biochem. Pharmacol. *33*:531
122. Denomme MA, Leece B, Li A, Towner R, Safe S (1986) Biochem. Pharmacol. *35*:277
123. Bannister R, Mason G, Kelley M, Safe S (1986) The effects of cytosolic receptor-modulation on the AHH-inducing activity of 2,3,7,8-TCDD. In: Proceedings, Dioxin 85 Symposium, Bayreuth, West Germany, Chemosphere
124. Tokarz RR, Harrison RW, Seaver SS (1979) J. Biol. Chem. *254*:9178
125. Leece B, Denomme MA, Li MA, Safe S unpublished results

Carcinogenic and Mutagenic Effects of PCBs

M. A. Hayes [1]

Polychlorinated biphenyls (PCBs) have been the subject of extensive experimental and epidemiological investigation as potential carcinogens. The available evidence indicates that PCBs are weak genotoxicants and initiators of carcinogenesis. However, at levels above a threshold, PCBs are potent promoters of hepatic carcinogenesis in laboratory rodents. While PCBs are able to enhance the biological generation of genotoxic metabolites of many carcinogens *in vitro,* they usually have the opposite effect *in vivo* and are capable of preventing carcinogenicity of various organ-specific carcinogens. Epidemiological studies on human and animal populations exposed to PCBs in the environment have so far not revealed clear evidence for carcinogenicity of PCBs under natural exposure circumstances. This review examines the available evidence for and against the view that environmental PCBs represent a significant potential carcinogen to humans. This evidence is interpreted on the basis of mechanistic analysis of the biological processes involved in chemical carcinogenesis. This analysis suggests that much of the evidence for potential carcinogenicity of PCBs in experimental systems leads to a substantial overestimate of the real risks to humans exposed to environmental levels of PCBs.

[1] Department of Pathology, University of Guelph, Guelph, Ontario, Canada, N1G 2W1

Environmental Toxin Series. Vol 1
© Springer-Verlag Berlin Heidelberg 1987

1 Introduction

Polychlorinated biphenyls (PCBs) were manufactured in large amounts for numerous industrial applications between the 1930s and mid 1970s and have become widely and persistently distributed throughout the environment (1–8). The manufacture and use of PCBs was curtailed during the 1970s because of accumulating evidence that PCBs had several real and potential harmful effects on human and animal populations exposed to them. Some small human populations have been accidentally or occupationally exposed to relatively high levels of commercial PCBs and developed well documented acute and chronic toxic sublethal effects (eg. "Yusho" and "Yu-Cheng") (9–14).

Although there has been little direct epidemiological evidence for harmful effects of low level exposure to PCBs from environmental contamination, PCBs were implicated as potential carcinogens on the basis of a range of experimental evidence obtained from laboratory animals exposed to them. This evidence for potential carcinogenicity lead to widespread concerns that PCBs in industrial waste stockpiles and in the environment represented a considerable unmeasured carcinogenic risk to human populations (15–17). These concerns still prevail, even though the environmental levels of PCBs are now decreasing (18) since the manufacture of PCBs was curtailed and since the implementation of more stringent measures to limit their discharge into the environment.

The long latent period for development of most neoplasms in humans, and the complexities and uncertainties in interpreting the significance of low-level PCB exposure as one of a range of concurrent uncharacterized carcinogenic risk factors, suggest that a direct epidemiological assessment of the carcinogenicity of environmental PCBs may be difficult to achieve in the near future (19, 20). As long as PCBs are considered carcinogenic, they are reasonably regarded as potentially noxious contaminants of our environment and are thus subject to various conservative regulatory policies consistent with such a judgement (21, 22). However, this cautious approach to such an unmeasured problem necessarily overestimates the predictive significance of available experimental evidence for carcinogenicity of environmental levels of PCBs.

In recent years, a clearer understanding of the mechanisms of carcinogenesis by PCBs and other chemicals has been emerging. Many earlier experimental studies demonstrated the potential for PCBs under some circumstances to increase the occurrence of neoplasms (23–31). Most of these experimental studies were not designed to measure carcinogenicity but rather to maximize the changes of detecting a potentiating effect of PCBs on the carcinogenic response. In the light of a clearer understanding of the mechanistic interrelationships among the multiple steps in chemical carcinogenesis, some of the experimental evidence and interpretations of potential carcinogenicity of PCBs should perhaps be re-evaluated. This review examines the accumulated evidence relating to actual and potential carcinogenicity of PCBs, with particular reference to its reliability in predicting carcinogenicity of low level exposure of humans and animals to PCBs in the environment.

2 Categories of Evidence for Carcinogenicity of PCBs

Available evidence relating to the carcinogenicity of PCBs can be arbitrarily considered in three main categories, namely (1) influences on mutagenicity and initiation of carcinogenesis, (2) influences on promotion and progression of carcinogenesis and (3) epidemiological evidence for carcinogenesis in naturally exposed populations. The available information included in categories 1 and 2 generally relates to studies directed at examining the influence of PCBs on selected stages or the complete sequence of a multistep carcinogenic process. The last category includes a limited number of equivocal studies of the rate of occurrence of cancers in naturally exposed animal and human populations. The majority of evidence comes from the experimental studies and its value depends on the operational quality of the various studies and the conceptual framework within which the studies have been designed and interpreted.

There are many concepts relating to how chemicals may cause neoplasms (32) and a thorough analysis of these is beyond the scope of this review. However, the data used for estimating carcinogenicity of PCBs need to be interpreted within a conceptual framework that is based on the actual biological effects and responses that occur during chemical carcinogenesis. For this reason, the demonstrable (ie not theoretical) biological mechanisms in chemical carcinogenesis *in vivo* will be briefly reviewed. Because PCBs have been mostly associated experimentally with hepatocarcinogenesis (23–31) and with increased hepatic activation of many other carcinogens (33, 34), most of this preliminary review of carcinogenic mechanisms pertains to the pathogenesis of cancer development in the liver.

3 Biology of Carcinogenic Responses to Chemicals

Numerous studies on the pathogenesis of neoplasias in various tissues exposed naturally or experimentally to chemical carcinogens have elucidated a number of critical biological steps or events that occur during the prolonged period of cancer development (35, 36). Many of these essential steps can be recognized by characteristic biological responses in target organs. The molecular mechanisms underlying some of these recognizable complex tissue responses are not yet well understood.

An understanding of the sequential relationship among the various carcinogenic responses has been developed from experimental models of chemical carcinogenesis in liver (35–37), skin (38, 39) and to a lesser extent in urinary bladder (35, 40), lung (41), mammary gland (42), and pancreas (43). Sequential carcinogenesis in these tissues has been conceptually considered to involve initiating, promoting and progression phases that are biologically and biochemically different (Fig. 1).

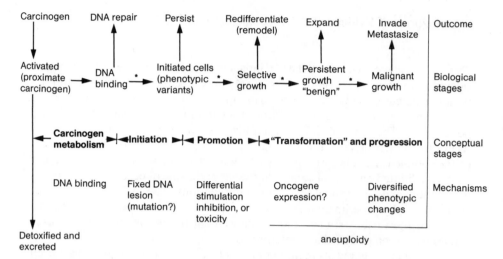

Fig. 1. Schematic representation of the biological, conceptual and mechanistic stages in chemical carcinogenesis. Asterisks indicate points at which cell proliferation is required

3.1 Initiation of Chemical Carcinogenesis

The first clearly recognizable tissue response is the generation of phenotypically altered cells that may have subtly different physiological functions (35–37, 44). These cells may have an altered growth response to various growth stimuli, such as those involved in developmental, regenerative or hyperplastic tissue growth. The appearance of phenotypically altered cells in a target tissue is considered a biological manifestation of the process of initiation of carcinogenesis (35, 37, 44, 45). For example, in chemically induced hepatocarcinogenesis in various laboratory rodents, known initiating carcinogens including nitrosamines, aromatic amines and polycyclic aromatic hydrocarbons all generate many small populations of phenotypically altered hepatocytes. These hepatocytes possess a variety of biochemical alterations including changes in levels of various enzymes involved in carbohydrate metabolism and xenobiotic detoxification (35, 37, 44–49). The emergence of these phenotypically altered "initiated" populations is a preneoplastic response that is directly and dose-dependently related to previous exposure to initiating dose of various carcinogens (44, 45). Moreover, these altered cells render the target tissue susceptible to the promoting influences of many other non-initiating chemical, nutritional or hormonal promoters (36, 44, 50), and are thereby a precursor to later cancer development.

Many strong initiating chemical carcinogens are genotoxic and capable of interacting covalently with DNA, either as direct acting chemicals such as N-methylnitrosourea, or indirect acting chemicals that must first be metabolized (e.g. by cytochromes P-450) to reactive derivatives (proximate carcinogens) (51, 52) (Fig. 2). The reactivity of carcinogens with DNA correlates with their mutagenicity in various *in vitro* assays such as the Ames' *Salmonella*-microsome assay (33, 34, 51, 52). The close correlation between mutagenicity in *in vitro* assays, and the

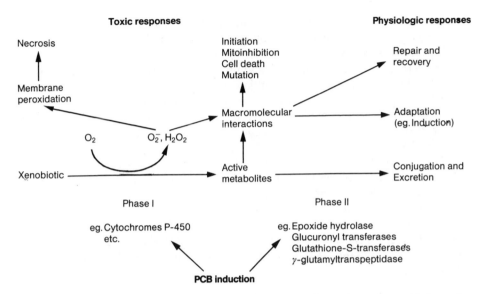

Fig. 2. Schematic representation of major biotransformation pathways for toxic xenobiotics and their responses

potential, under appropriate conditions, to generate phenotypically altered populations of cells *in vivo* suggests that phenotypic alterations might involve mutagenic lesions. However, there is some evidence that many altered populations of hepatocytes may later reacquire a normal phenotype (44, 53), suggesting that the emergence of phenotypically altered cells in response to genotoxic carcinogens is not necessarily a consequence of permanent mutagenic changes to primary DNA sequences (36). For example, non mutagenic mechanisms such as transient loss of DNA methylation patterns involved in regulating phenotypic expression might occur after alkylated DNA has been repaired. The reversibility of phenotypic alteration under some conditions (44) leaves as yet an unresolved question concerning the concept of permanence of initiaton.

Nevertheless, it is clear that the propensity of initiating carcinogens to damage DNA is an important aspect of the initiation process, regardless of the actual molecular mechanisms and conceptual principles involved. Thus mutagenicity in various test systems is probably predictive of the potential of a chemical to initiate because mutation and initiation each involve primary chemical injury to DNA that is incompletely repaired before DNA replication.

An important aspect of the initiation process is the requirement for cell proliferation. Unless carcinogen-injured cells are stimulated to undergo a round of DNA replication before the DNA lesions are fully repaired, initiation of carcinogenesis does not occur (36, 37, 46, 54–56). Thus, covalent binding of genotoxic metabolites to DNA does not necessarily lead to initiation (46, 56).

For example, in adult rat liver, cell proliferation is normally very low so this organ can withstand and repair a degree of DNA damage by carcinogens such as benzo(a)pyrene, N-methylnitrosourea or 2-acetylaminofluorene (2-AAF) without the generation of phenotypically altered initiated cells (46, 50, 54–56). However, under the influence of a cell proliferative stimulus such as developmental liver growth in juvenile rats (37), or regenerative hepatocellular proliferation after necrosis (56) or partial hepatectomy (PH) (46, 54), these and many other genotoxic carcinogens are potent initiators. However, single doses of genotoxic hepatocarcinogens such as aflatoxin B_1 and pyrrolizidine alkaloids that also strongly inhibit hepatocellular proliferation are weak initiators, presumably because they prevent carcinogen damaged hepatocytes from proliferating before their DNA is adequately repaired (57, 58).

This aspect of the initiation process *in vivo* is important because it indicates a discontinuity between the events involved in DNA binding and damage by genotoxic carcinogens, and the subsequent initiation process. Thus mutagenicity or genotoxicity in isolated *in vitro* systems is not necessarily predictive of the capacity of a chemical to initiate carcinogenesis in a particular organ *in vivo*. This is especially true for the majority of short-term assay systems that lack normal levels of protective mechanisms such as those involved in chemical detoxification or DNA repair.

3.2 Promotion in Chemical Carcinogenesis

A promoter of carcinogenesis is a chemical or other influence that accelerates the development of cancer in a tissue that has previously been initiated by a genotoxic insult and cell proliferation. During promotion, initiated cells are selectively stimulated to grow into phenotypically distinct preneoplastic populations such as papillomas, polyps, plaques and nodules (35, 44). The selective enlargement of preneoplastic lesions is eventually associated with an increased likelihood of altered cells undergoing further poorly understood changes that result in persistent proliferation in the absence of the promoter.

Most tumor promoters induce altered differentiation in target organs, and also induce phenotypic changes in initiated cells that otherwise might be indistinguishable from their surroundings (44). Some so called "pure" promoters do not themselves initiate (38, 59, 60), but many initiating carcinogens may have promoting influences (47, 57, 59, 61, 62). Unlike the process of initiation, which can occur as a brief insult and an apparently long lasting or perhaps permanent change in phenotypic control (35, 36, 38), many promoting influences require a prolonged effect that is apparently reversible when the promoter is removed (38, 44, 49, 53). However, it is presently unclear if the reversibility applies to the phenotypic changes induced in initiated cells by promoters, or to their predisposition to develop into cancers.

Promotion of preneoplastic growth may occur by several different biological mechanisms. For example, xenobiotics that induce hyperplastic or hypertrophic growth may selectively stimulate growth of some phenotypically altered cells that are more responsive to these stimuli (35, 44). Included in this group are promoters such as phorbol esters, PCBs, barbiturates, and trophic or anabolic hormonal agents. Alternatively, xenobiotics or nutritional insults that inhibit the survival or proliferation of normal cells can selectively permit proliferation of those phenotypically altered initiated cells resistant to these insults (45, 47, 57, 63). In-cluded in this latter category are a wide range of cytotoxic or mitoinhibitory toxi-cants, such as non initiating necrogens (64) and also many genotoxic carcinogens at dosages that are differentially more toxic to non initiated cells (50, 57, 58). A large proportion of initiated hepatocyte populations contain elevated levels of various enzymes including glutathione-S-transferases and UDP-glucuronyltrans-ferases that are involved in detoxification of many xenobiotics (44, 47, 65). The increased levels of glutathione-dependent and other cytoprotective mechanisms might also enhance the ability to altered cells to withstand a cytocidal response after covalent molecular binding by a carcinogen has occurred (56). Promoted populations of initiated hepatocytes also have reduced levels of various cy-tochrome P-450 dependent activities and so are less able to generate toxic meta-bolities or cytotoxic oxygen radicals (44, 47, 65). Thus, in the resistant-hepatocyte models of chemical hepatocarcinogenesis, agents such as 2-AAF and pyrrol-izidine alkaloids that inhibit normal hepatocytes from proliferating are strong promoters of growth of resistant altered cells when the liver is subjected to a re-generative growth stimulus such as necrosis or PH (35, 45, 50, 57).

3.3 Progression in Chemical Carcinogenesis

When initiated cells have been promoted, some may undergo further changes that result in persistent proliferation (35, 36, 38, 66). These cells eventually become in-dependent of the promoting influence and substantially less dependent on normal growth regulatory stimuli (67, 68). The molecular events involved in these later events that are perhaps equivalent to "transformation" are poorly defined but might include expression of various oncogenes. Xenobiotics can enhance the pro-cesses leading to persistent proliferation and progression of cancers. For example, genotoxic carcinogens increase the conversion of skin papillomas to carcinomas (61). However, there is presently little experimental data to indicate how the mo-lecular mechanisms of progression might be modulated.

The influence of xenobiotics on subsequent events involved in diversification and aggressive growth behaviour (ie local invasion or metastasis) have not been adequately evaluated. Most hepatocellular carcinomas that develop from resis-tant preneoplastic nodules maintain many of the early phenotypic alterations that rendered them resistant to chemical injury (65, 66). Persistently proliferating he-patocytes in nodules do not progress more rapidly when exposed to promoters such as phenobarbital or PCBs (69–71) but their progression is affected by some hormones (72). Since exogenous agents are not required for persistently pro-liferating nodules to progress to cancers, this stage of the carcinogenic process has been considered to be "self-generating" (47). A corollary hypothesis is that pro-moters increase the pool of cells that randomly undergo these undefined later events, suggesting that progression is a stochastic process rather than a specific molecular insult by xenobiotics.

4 Carcinogenicity of PCBs

4.1 Influences on Initiation of Carcinogenesis

Virtually all isomers and congeners of PCBs are remarkably stable chemicals that are not readily converted by biotransformation enzymes into reactive intermediates that could potentially damage DNA (7). This is consistent with observations that short-term exposures of rodents to PCBs generate very few or no initiated hepatocytes in various model systems (59, 73, 74) nor do they initiate carcinogenesis in the skin of mice (75). Even when administered to rats undergoing induced or developmental hepatocellular proliferation, various complex mixtures and several pure PCBs failed to initiate in the liver (74). Even 2,2′,5,5′-tetrachlorobiphenyl, which can be activated to arene oxides and covalently bind to cellular macromolecules, does not initiate under these conditions (74). There is evidence that some PCBs can covalently bind to DNA and cause genotoxic effects (eg chromosomal breaks) in some in vitro systems (76–80). However, the inactivity of PCBs as initiators in in vivo assays correlates with their failure to cause mutations in various short-term in vitro assays for genotoxicity (81, 82).

This collected evidence indicates that short-term exposures to various pure PCBs and also commercial mixtures of PCBs that may contain various chlorinated dibenzofurans and dioxins, are, at the worst, very weak initiators of carcinogenesis. However, mice and rats exposed continuously to PCBs for periods of 2 years or more develop preneoplastic liver (23–28, 84–88) and gastric lesions (89, 90) and also a low incidence of frank hepatocellular carcinomas and gastric carcinomas. This suggests that initiation may occur under some conditions of long term exposure to PCBs (90). This does not necessarily mean that the PCBs themselves are responsible for the initiation process because there is evidence that some strains of rats and mice "spontaneously" develop phenotypically altered "initiated" populations of hepatocytes later in life (44, 71, 91). Thus the carcinogenicity of the prolonged exposures to PCBs in these studies could be a reflection of the promoting influence of PCBs on preneoplastic populations initiated by other, as yet unidentified, mechanisms.

4.2 Influences on Initiation and Mutagenicity by Other Genotoxic Carcinogens

PCBs and many other halogenated hydrocarbons are potent inducers of various hepatic microsomal cytochrome P-450 isozymes (7, 92). The cytochrome P-450-dependent monooxygenases are perhaps the most important mechanisms by which precarcinogens are activated to reactive intermediates and proximate carcinogens (51, 52). Thus, it might be anticipated that hepatocytes previously induced by PCBs would be more susceptible to initiation by many carcinogens that require microsomal activation for genotoxicity (93). Indeed, numerous carcinogens have been found to have enhanced mutagenicity in the Ames test when microsomal or post-mitochondrial supernatant subfractions of PCB-induced livers are coadministered with the carcinogens (33, 34). For this reason, the sensitivity of many *in vitro* screening assays for genotoxic carcinogenicity is increased by the routine inclusion of microsomal preparation from PCB-pretreated rats or other animals (33, 34, 93).

It is beyond the scope of this review to catalogue all of the numerous carcinogens for which genotoxicity is potentiated *in vitro* by PCBs, but comprehensive lists of some of these carcinogens are already available (34, 93). While this collective evidence has been frequently used to implicate PCBs as important "co-carcinogens" (ie potentiators of carcinogenesis by other agents), the evidence indicating that PCBs potentiate initiation by other carcinogens is limited. Under some conditions, PCBs have been found to slightly increase the initiating activity of benzo(a)pyrene (94) and nitrosamines (95, 96). However, other studies have shown that PCBs markedly diminish the carcinogenicity dimethylnitrosamine, diethylnitrosamine (95, 97), azo dyes, 2-acetylaminofluorene (98), and aflatoxin B_1 (99, 100), all of which are rendered more mutagenic in the Ames test by PCB-induced microsomes.

There are several possible reasons for these contradictory influences of PCBs on mutagenicity *in vitro* and initiation *in vivo*. Hepatic microsomal subcellular fractions metabolize carcinogens differently from intact hepatocytes possibly because subcellular spatial organization of organelles and excretion mechanisms are abnormal (101). Microsomal preparations appear to be functionally deficient in various phase II detoxification systems, including those dependent on a constant supply of substrates and cofactors for conjugation, such as glutathione and UDP-glucuronic acid (102). However, PCBs induce many phase II functional detoxification systems *in vivo* (103–106). Also, the *Salmonella* strains used in the Ames tests are deficient in various normal DNA repair activities (33), whereas intact mammalian cells are fully functional in this important protective mechanism against initiation. Thus, the capacity of PCBs to potentiate covalent binding and genotoxicity of various carcinogens in systems that employ subcellular fractions as activation systems is not necessarily a good estimate of their ability to initiate carcinogenesis in intact animals or cells. In spite of this clear discrepancy, the fact that PCBs can potentiate mutagenicity of other agents under contrived circumstances that do not represent the true functional status of normal cells, is frequently cited as evidence that PCBs represent a potential enhancer of carcinogenicity of other xenobiotics.

PCB pre-exposures can however potentiate some other toxic effects of xenobiotics. For example, liver cells of rats previously induced by some PCBs are more susceptible to the necrotizing toxicity of high concentrations of bromobenzene, acetaminophen or the hepatocarcinogen 2-AAF (107, 108). While these potentiating effects on lethal toxicity reflect an increased activation of the xenobiotics by PCB-induced cytochrome P-450 species (106, 107), the important distinction to be made in these situations is that hepatocellular necrosis only occurs when the cells have exhausted the relevant substrates for either the detoxification of the metabolites generated (109) or the protection of the cell against oxidant injury (110). Thus, this situation is somewhat analogous to the potentiation of mutagenicity in the systems that employ subcellular fractions functionally deficient in phase II detoxification pathways. By comparison, many genotoxic carcinogens that initiate at relatively low sublethal doses are unlikely to saturate the available detoxification and cytoprotection systems. In fact, many phase II detoxification functions are induced by PCBs including epoxide hydrolase, glutathione-S-transferases, UDP-glucuronyl transferases and DT-diaphorase (quinone reductase) (103–106). All of these enzymes might potentially reduce the covalent binding and initiating activity of low concentrations of carcinogens, whose reactive metabolites are substrates for the induced detoxification enzymes.

This explanation is consistent with the observations that PCB pretreatments diminish initiating carcinogenicity of many carcinogens while at the same time enhancing their metabolism (95–100, 108). The further possibility that PCBs might also enhance DNA repair mechanisms has not yet been adequately evaluated. An important conclusion suggested by these various observations is that low levels of environmental PCBs appear more likely to prevent than enhance initiation by other environmental xenobiotics. This conclusion needs to be more comprehensively substantiated by experimental and epidemiological data. Meanwhile, the opposite conclusion based on the *in vitro* test data alone can reasonably be considered an exaggerated estimate of the potential for low levels of environmental PCBs to act as cocarcinogens by potentiating genotoxicity.

4.3 Influences on Promotion and Progression of Carcinogenesis

Many studies in laboratory rodents previously initiated with various genotoxic carcinogens have clearly established that subsequent exposure to PCBs promotes carcinogenesis in the liver (23–25, 29–31, 59, 73, 94, 95, 111). In these studies, PCBs increase the numbers of phenotypically altered populations of hepatocytes and accelerate their rate of development into persistent nodules, hepatomas and carcinomas.

Promotion by PCBs is dose-dependent and there appears to be a threshold dose below which promotion of preneoplastic liver lesions is not observed (73). This threshold for promotion may be well above the levels encountered in animals and humans exposed to environmental PCBs. Female rats appear to be more susceptible than males to hepatic tumor promotion by PCBs (25, 27, 89, 94) but the reasons for and significance of this difference in sensitivity between sexes are unknown. Promotion of liver carcinogenesis by PCBs appears to be biologically analogous to promotion by other inducing xenobiotics such as phenobarbital (44, 60, 62, 90, 91, 112). Promotion by phenobarbital occurs only after prolonged exposure (60, 62) and may be reversible after exposure is terminated (44). However, for PCBs that persist in body tissues for long periods after single exposures (73, 84), repeated exposure might not be required for promotion (113), and it is less likely that the promoting influence would subside after a brief exposure to PCBs.

The promoting influences of PCBs on carcinogenicity in rodents appears largely specific for the liver. PCBs do not promote carcinogenesis in the bladder or skin previously initiated by genotoxic agents with demonstrable initiating carcinogenicity in bladder and skin respectively (75, 114). Although PCBs stimulate proliferation of cells in the thyroid gland (115, 116), the possibility that they promote thyroid carcinogenesis in animals exposed to initiating thyroid carcinogens has not been adequately investigated. However, phenobarbital promotes thyroid carcinogenesis in rats previously initiated with N-methylnitrosourea (117) or N-bis-(2-hydroxypropyl)nitrosamine (118).

Promotion of hepatocarcinogenesis by PCBs correlates closely with but not consistently with their ability to induce microsomal mixed function oxygenases in rodent liver (44, 73). The relationship between induction and promotion is unclear, especially since PCBs evidently induce various microsomal enzymes in many non hepatic tissues in which they do not appear to promote, including skin (75), kidney (119), testis (119), placenta (120, 121), lung (119, 120, 122), and intestine (123). Hepatocarcinogenic promoting regimes of PCBs apparently do not enhance renal carcinogenesis by nitrosamines (Hayes, M.A. unpublished observations). Also, regimens of Aroclor 1254 that promote hepatocarcinogenesis in mice initiated with dimethylnitrosamine (DMN) do not promote the pulmonary adenomas that are also initiated by DMN (95). While these differences in organ specificity for promotion by PCBs might perhaps be explained by the different induction profiles for various target organs, it presently is unclear if the induction response is essential to the promotion process. The lack of data from long term studies in animals subjected to various initiators of non hepatic tissues indicate that it is premature to conclude that PCB promotion is a phenomenon restricted to hepatocarcinogenesis in rodents.

Gastric carcinogenesis by chronic Aroclor 1254 administration (89) has been considered a result of promotion rather than initiation (90). On the assumption that PCBs are virtually inactive as initiators, the pathogenesis of gastric neoplasms in response to Aroclor 1254 can be explained by a promoting influence on preneoplastic lesions initiated either spontaneously or by chemicals such as unidentified components in the experimental diets used in these studies. PCBs have been recognized to stimulate mucosal hypertrophy in primates and rodents (90, 124, 125) and such a tissue response could promote by growth stimulation over a prolonged period. However, a more definitive analysis of this possibility awaits further long-term studies in animals exposed to PCBs after administration of known initiators of gastric carcinogenesis such as N-methyl-N-nitrosoguanidine.

The biological mechanisms by which PCBs and other inducers promote carcinogenesis are unknown but several hypotheses have been suggested. PCBs functionally resemble many other tumor promoting xenobiotics including other halogenated hydrocarbons, barbiturates, steroid hormones (44, 91, 126–128) hypolipidemic drugs and phthalate esters (91, 129). Most of these agents stimulate hepatocellular hypertrophy in rodents and increase microsomal (44, 126) or peroxixomal (129) components. However, these agents are all potent stimulators of hepatocellular proliferation (hyperplasia) (44, 126). The hyperplastic effect of these inducers could promote preneoplastic growth according to the concept of promotion by differential stimulation (35, 36, 44, 47).

Other biological mechanisms for the promoting effect of inducing xenobiotics are also supported by experimental evidence. Barbiturates have been shown to prevent remodelling by stabilizing the altered phenotype of initiated preneoplastic liver lesions (44, 130). This phenomenon may explain the enhanced survival of transplanted altered cells in phenobarbital-treated rats (69–71). Enhanced survival would increase the size of initiated populations and increase the likelihood of subsequent events in the preneoplastic lesions that are prevented from remodelling to the normal hepatocyte phenotype. Another phenomenon that may be relevant to the biological mechanism by which inducing xenobiotics promote and enhance survival of altered cell is by inhibition of apoptosis (91). Apoptosis is a mode of cell death that is termed "programmed cell death" involved in the involution of surplus growth after hepatic hypertrophy (44) and in the rapid turnover of hepatocytes in persistently proliferating preneoplastic nodules (131). Various inducing xenobiotics could contribute to an excessive accumulation of normal and preneoplastic hepatocytes that are inhibited from undergoing apoptosis.

A further biological mechanism by which PCBs might promote is by toxic injury to non initiated hepatocytes. If the altered phenotype of initiated cells renders them less susceptible to toxic influences of PCBs that affect longevity (survival) or proliferative capacity of non initiated cells, then PCBs might promote according to the hypothesis of differential inhibition (35, 36). This possibility has not been adequately investigated. Chronic high dose PCB toxicity is associated with adenofibrosis and bile duct proliferation (27, 83, 84), both of which are liver responses characteristic of numerous toxic and mitoinhibitory xenobiotics that promote hepatocarcinogenesis by differential inhibition (35, 36, 57, 58, 108). Although promoting regimens of phenobarbital do not affect non-initiated hepatocyte survival during the early stages of promotion of hepatocarcinogenesis in rats (70), it is not yet known if inducing promoters affect the long-term survival and proliferative capacity of non initiated hepatocytes in carcinogen-damaged livers.

These various biological mechanisms explain how PCBs might promote the development and expansion of preneoplastic lesions. Little is known about the biological influences of PCBs and inducing xenobiotics on the progression phases of cancer development (ie the emergence and diversification of cancers in persistently proliferating nodules). Barbiturates and Aroclor 1254 have little influence on the progression of transplanted persistent liver nodules in rats (70) suggesting that the essential later events in progression of carcinogenesis may be independent of the influences of an exogenous liver promoter.

These various possible biological mechanisms of promotion by PCBs generally reflect complex physiological tissue responses. Since these responses are poorly understood in mechanistic terms, it is perhaps premature to speculate on the molecular mechanisms involved in PCB promotion. Some authors have proposed that such mechanisms be termed "epigenetic", to distinguish them from the better characterized "genotoxic" mechanisms involved in initiation events by carcinogens that damage DNA (132). While such a classification is useful operationally, it does not define the molecular mechanisms by which inducing xenobiotics promote liver cancer. There is some reason to suspect that the promoting xenobiotics activate a cytoplasmic receptor-mediated adaptive cell response analogous to that induced by many steroid hormones (7).

In addition to having a promoting influence on previously initiated liver cells, PCBs are also capable of preventing promotion. PCBs and other xenobiotic inducers have a well recognized capacity to prevent promotion by toxic inhibitors of hepatocellular proliferation, such as 2-AAF, aflatoxin B_1 and azo dyes (25, 60, 99–101, 108, 133–136). These hepatotoxins are potent liver carcinogens largely because they inhibit normal hepatocellular regeneration, and thus selectively stimulate (promote) initiated hepatocytes with the resistant initiated phenotype (47, 74, 108). As such, this mechanism of promotion by toxic xenobiotics differs biologically from the promoting mechanism involved in the adaptive proliferative response to inducing xenobiotics. An important aspect of these two different mechanisms of promotion is that they may negate each other under some conditions. For example, in rats previously initiated with diethylnitrosamine, low doses of various PCBs with different cytochrome P-450 induction capacities abrogate the ability of 2-AAF to promote resistant liver nodule growth (108). Other inducers such as phenobarbital and 3-methylcholanthrene similarly inhibit carcinogenesis by 2-AAF, aflatoxin B_1 and nitrosamines (60, 133–136). Although the different inducers frequently have opposite influences on the rate of cytochrome P-450-dependent generation of genotoxic metabolites of 2-AAF (108, 137), they all similarly protect against its mitoinhibitory toxicity (108). PCB-induced phase II detoxification systems probably prevent toxic inhibition of cell proliferation and more than compensate for the increased amounts of active metabolites generated by different induced cytochromes P-450 (103–106).

The possible significance of promoting and antipromoting influences of PCBs on hepatocarcinogenesis in humans cannot readily be extrapolated from studies conducted in laboratory rodents. It is possible that rodents are unusually susceptible to hepatic tumor promotion by inducing xenobiotics, (138, 139) possibly because they spontaneously generate numerous preneoplastic foci and nodules as they get old (44, 71, 91). While this would likely render rats susceptible to promotion by PCBs and other inducing xenobiotics, humans apparently do not develop these lesions to anywhere near the same extent. Long-term exposure of epileptics to phenobarbital does not result in an increased rate of preneoplastic or neoplastic liver lesions (140, 141). Also, long-term surveillance of human populations exposed to other rodent liver promoters such as dioxins (142) has not revealed an increased susceptibility to liver cancer. Accordingly, some authors have considered that promotion of hepatocarcinogenesis by inducing xenobiotics in overfed, sexually inactive laboratory rats exosed to high levels for their entire life is not representative of the susceptibility of humans to these agents in the environment (20, 94, 138, 139). There remains a possibility that PCBs might promote hepatocarcinogenesis in humans whose livers have previously sustained substantial genotoxic injury and initiation, but the prevalence of such a potential predisposition is not known. In spite of many studies on body and adipose tissue burdens of humans exposed to PCBs from environmental or occupational sources (9–14, 143–147), there is very little information available to indicate the extent to which human liver undergoes the adaptive enzyme induction response to PCBs observed in liver and other tissues of rodents (121). However, human liver is induced by phenobarbital (126) suggesting that the induction response might render human liver less susceptible to the initiating or promoting influences of other cytotoxic and genotoxic xenobiotics.

The important conclusion from these observations is that the role of PCBs as promotors of human liver carcinogenesis cannot be reliably predicted from their influences on metabolic activation of carcinogens or on promotion of carcinogen-initiated livers in rodents. Under natural circumstances, pre-exposure or concurrent exposure of human to low levels of inducing xenobiotics such as PCBs might protect against carcinogenesis by other compounds such as aflatoxin B_1 and nitrosamines that are activated by PCB-induced microsomes to initiating or cytotoxic metabolites. In contrast, PCBs might not promote human liver carcinogenesis unless they are present at high levels in tissues of humans that have previously been exposed to strong initiating carcinogens. Thus designating PCBs as promoters on the basis of contrived initiation-promotion models does not appropriately acknowledge the anti-promoting effects of low levels of PCBs that precede or coincide with exposure to other toxic promoting and/or initiating carcinogens.

5 Epidemiology of Neoplastic Disease in PCB-Exposed Humans and Animals

In spite of the widespread concerns about the potential carcinogenicity of PCBs in populations exposed to them occupationally, accidentally or environmentally, there is virtually no clear epidemiological evidence to substantiate these concerns (19). Long-term monitoring of "Yusho" victims (11) and workers involved in the manufacture or use of transformers and other devices containing PCBs (148, 149) has largely failed to demonstrate that apparently increased rates of neoplastic disease can specifically be attributed to PCB exposure. Similarly, the levels of PCBs in human tissues do not correlate with the occurrence of specific cancers such as breast cancers (147). There are, however, many confounding aspects to any epidemiological analysis of the cause-effect relationships between chemicals and diseases such as cancers that have a long latent period for detection. Numerous other possible carcinogens (10) and other risk factors may be more important than the PCBs themselves. Under such conditions, lack of evidence for a clear association between a putative risk factor and cancer does not necessarily constitute evidence for the conclusion that the risk factor is insignificant.

In recent years, some epidemiological studies have been directed toward evaluating the relationship between environmental pollution and the occurrence of neoplasms in wild fish populations. There is strong evidence that some bottom-dwelling fresh water fish in polluted aquatic environments have a greatly increased rate of skin and hepatic neoplasms (150–152). These fish have higher body burdens of PCBs but also have increased rates of exposure to a wide range of potentially carcinogenic pollutants including polycyclic aromatic hydrocarbons and heavy metals (151, 152). This complex situation suggests that pollution in general is causally related to neoplastic development in bottom-dwelling fish. However, PCBs persist in body fats and bioaccumulate in all living organisms in the aquatic ecosystem. Predatory fish at the top of the food chain have substantially greater body burdens of persistent PCBs and other halogenated hydrocarbons (152). However, the fact that these fish with highest body burdens have much lower rates of neoplasia suggests that the persistent chemicals are not the major risk factor for carcinogenesis in fish in polluted aquatic environments (152). The dissociation between persistence in biological tissues and the potential for carcinogenicity is consistent with the biochemical evidence that carcinogenic and toxic chemicals are reactive in biological systems and thus less likely to accumulate. Such a concept has important implications because it suggests that highly chlorinated PCB isomers and congeners that persist in the environment and in fish (155) are less likely to be carcinogenic than the PCBs and other pollutants in water sediments.

Further multifactorial epidemiological studies on the occurrence of naturally occurring cancers in humans, fish and other animals are clearly required. Such studies may eventually determine if PCB levels in tissues are causally associated with carcinogenesis or are merely a readily detectable indication of general exposure to many industrial pollutants. PCBs will remain etiological candidates for pollution-associated cancers until either causally linked with or dissociated from their occurrence by sound epidemiological analyses.

6 Conclusions

Concerns that environmental PCBs represent a potential carcinogenic risk to humans and animals are frequently expressed, but are largely based on experimental evidence that is perhaps of equivocal predictive significance. PCBs appear to be at the worst very weak genotoxicants or initiators of carcinogenesis in various systems. Their well established activity at moderately high levels as promoters of hepatocarcinogenesis in rodents should not necesarily be accepted as clear predictive evidence for a similar effect activity in humans. Many xenobiotics with no known hepatocarcinogenic in humans are, like PCBs, strong promoters of liver tumor growth in rodents that appear inherently highly susceptible to this response. Evidence that PCBs enhance genotoxicity and mutagenicity of many other xenobiotics in various *in vitro* test systems is in direct contrast to their protective role against carcinogenicity of many genotoxic carcinogens *in vivo*. Furthermore, PCBs should not be universally regarded as liver tumor promoters because they strongly prevent the promoting activity of other environmental carcinogens such as aflatoxin B_1.

Collectively, this evidence suggests that PCBs may potentially be carcinogenic under some specific conditions. However, under natural exposure circumstances, PCBs are perhaps more likely to prevent carcinogenesis than enhance it. Accordingly, the risk estimates based on the worst-case analysis of the potentially carcinogenic effects are likely to be substantial overestimates of the real risks.

The lack of clear epidemiological evidence for carcinogenicity of environmental PCBs in humans and animals is apparently consistent with this mechanistic interpretation of the various pieces of experimental evidence. However, the number of thorough epidemiological studies conducted is still quite small. More discriminating studies of PCBs and other potential carcinogenic risk factors are necessary before it would be reasonable to conclude that PCBs are not human carcinogens. Similarly, additional experimental evaluation is required to explore the possibility that PCBs might promote carcinogenesis in tissues other than liver in animals exposed to various tissue-specific initiating agents. Such a mechanistic evaluation is more likely to provide a sound conceptual basis for regulatory policies than is the continued demonstration of carcinogenic or co-carcinogenic potential under contrived, highly sensitive experimental conditions.

PCBs are clearly toxic and hazardous to humans exposed accidentally to high levels. While this fact alone warrants considerable precautions to minimize human exposure to PCBs, it should not necessarily serve as a basis for extrapolation in defining the hazards associated with low-level exposure. Even in those individuals accidentally exposed to high levels of PCBs there is no clear documented evidence for lethal or carcinogenic toxicity. A cursory perspective of the environmental PCB problem tends to support a view that all exposure should be minimized to avoid any real or imagined risk. However, such a perspective does not acknowledge the possibility that low-level exposure to PCBs could even be beneficial, for example by enhancing various detoxification enzyme systems.

A broader view of the PCB problem should include an analysis of all risks and benefits associated with the use and elimination of PCBs. It appears reasonable to consider the benefits to human health that have occurred through industrial use of PCBs, such as in the prevention of accidents from electrical failures and fires. Conversely, the risks to humans engaged in hazardous waste storage, containment and destruction may in fact be greater than the risks to humans exposed to PCBs diluted in the environment. Another factor that warrants inclusion in the total risk-benefit analysis of the PCB problem is the social and economic cost of diverting large amounts of human and financial resources toward management and analysis of the PCB problem. This diversion of resources must inevitably curtail the ability to recognize and manage other more significant environmental or health problems.

7 References

1. Jensen S (1966) New Scientist. *32*:612
2. Risebrough RW, Rieche P, Herman SG, Peakall DB, Kirven MN (1968) Nature. London *220*:1098
3. Price HA, Welch RL (1972) Environ. Health Perspect. *1*:73
4. Wasserman M, Wasserman D, Cucos S, Miller HJ (1979) Ann. N.Y. Acad. Sci. *320*:69
5. Brinkman UATh, de Kok A (1980) In: Kimbrough RD (ed) Halogenated Biphenyls, Terphenyls, Naphthalenes, Dibenzodioxins and Related Products. Elsevier/North Holland, Amsterdam, p 1
6. Atlas E, Giam CS (1981) Science. *211*:163
7. Safe S (1984) CRC Crit. Rev. Toxicol. *13*:319
8. Tiernan JO, Taylor ML, Garrett JH, Van Ness GF, Solch JG, Wagel DJ, Ferguson GL, Schecter A (1985) Environ. Health Perspect. *59*:143
9. Kikuchi M, Masuda Y (1976) In: Higuchi K (ed) PCB Poisoning and Pollution. Academic Press, New York, p 69
10. Masuda Y, Yoshimura H (1984) Am. J. Indust. Med. *5*:31
11. Kikuchi M (1984) Am. J. Indust. Med. *5*:19
12. Urabe H, Asahi M (1985) Environ. Health Perspect. *59*:11
13. Hsu S-T, Ma C-I, Hsu SK-H, Wu S-S, Hsu NH-M, Yeh C-C (1984) Am. J. Indust. Med. *5*:71

14. Chen P, Wong C-K, Rappe C, Nygren M (1985) Environ. Health Perspect. *60*:59
15. Allen JR, Norback DH (1977) In: Hiatt HH, Watson JD, Winsten JA (eds) Origins of Human Cancer. Vol. 4. Cold Spring Harbor, New York, p 173
16. Kimbrough R, Buckley J, Fishbein L, Flamm G, Kasza L, Marcus W, Shibko S, Teske R (1978) Environ. Health Perspect. *24*:173
17. DiCarlo FJ, Seifter J, DeCarlo VJ (1978) Environ. Health Perspect. *23*:351
18. Kee N (1983) Nature. London *303*:653
19. Kimbrough RD (1985) Environ. Health Perspect. *59*:99
20. Griem H, Wolff T (1984) ACS Monogr. *182*:525
21. Somers E (1979) Can. J. Publ. Health *70*:388
22. Cordle F (1984) Regul. Pharmacol. Toxicol. *4*:236
23. Ito N, Nagasaki H, Arai M, Makiura S, Sugihara S, Hirao K (1973) J. Natl. Cancer Instit. *51*:1637
24. Nishizumi M (1976) Cancer Lett. *2*:11
25. Kimura NT, Kanematsu T, Baba T (1976) Z. Krebsforsch. *87*:257
26. Kimbrough RD, Linder RE (1974) J. Natl. Cancer Instit. *53*:547
27. Kimbrough RD, Squire RA, Linder RE, Strandberg JD, Montali RJ, Burse VW (1975) J. Natl. Cancer Instit. *55*:1453
28. National Cancer Institute, Bethesda, MD (1977) CAS No. 27323-18-8 National Technical Information Service, Springfield Virginia, USA PB-279, 624/IGA, NCI-CG-TR-38 DHEW/Publ/NIH-78-838
29. Tatematsu M, Nakanishi K, Murasaki G, Miyata Y, Hirose M, Ito N (1979) J. Natl. Cancer Instit. *63*:1411
30. Ito N, Tatematsu M, Nakanishi K, Hasegawa R, Takano T, Imaida K (1980) Gann. *71*:580
31. Preston BD, Van Miller JP, Moore RW, Allen JR (1981) J. Natl. Cancer Instit. *66*:509
32. Becker F (1982) Cancer: A Comprehensive Treatise. 2nd Edit. Vol. 1, Plenum Co., New York
33. Ames BN, McCann J, Yamasaki E (1975) Mutat. Res. *31*:347
34. de Serres FJ, Ashby J (1981) Evaluation of Short-term Tests for Carcinogens: Report of the International Collaborative Program. Elsevier/North Holland
35. Farber E, Cameron RG (1980) Adv. Cancer Res. *31*:125
36. Farber E (1984) Cancer Res. *44*:4217
37. Pitot HC, Sirica AE (1980) Biochem. Biophys. Acta. *605*:149
38. Boutwell RK (1974) CRC Crit Rev. Toxicol. *2*:419
39. Slaga TJ, Fischer SM, Nelson K, Gleason GA (1980) Proc. Natl. Acad. Sci. USA *77*:659
40. Hicks RM, Chowaniec J (1978) Int. Rev. Exp. Pathol. *18*:199
41. Witschi HR, Morse CC (1983) J. Natl. Cancer Instit. *71*:859
42. Yokoro K, Nakano M, Ito A, Nagao K, Kodama Y, Hamada K (1977) J. Natl. Cancer Instit. *58*:1777
43. Longnecker DS, Crawford BG (1975) J. Natl. Cancer Instit. *35*:2249
44. Schulte-Hermann R (1985) Arch. Toxicol. *57*:147
45. Solt DB, Farber E (1976) Nature, London *263*:701
46. Tsuda H, Lee G, Farber E (1980) Cancer Res. *40*:1157
47. Farber E (1984) Cancer Res. *44*:5463
48. Moore MA, Hacker H-J, Bannasch P (1983) Carcinogenesis. *4*:595
49. Enomoto K, Farber E (1982) Cancer Res. *42*:2330
50. Solt DB, Cayama E, Tsuda H, Enomoto K, Lee G, Farber E (1983) Cancer Res. *43*:188
51. Miller EC (1978) Cancer Res. *38*:1479
52. Rajalakshmi S, Rao P, Sarma DSR (1982) In: Becker F (ed) Cancer: A Comprehensive Treatise. Vol. 1. 2nd Edit. Plenum Co., New York, p 335
53. Tatematsu M, Nagamine Y, Farber E (1983) Cancer Res. *43*:5049
54. Cayama E, Tsuda H, Sarma DSR, Farber E (1978) Nature, London *275*:60
55. Columbano A, Rajalakshmi S, Sarma DSR (1981) Cancer Res. *41*:2079
56. Ying TS, Sarma DSR, Farber E (1981) Cancer Res. *41*:2096
57. Hayes MA, Roberts E, Farber E (1985) Cancer Res. *45*:3726
58. Neal GE, Butler WH (1978) Br. J. Cancer *37*:55

59. Tatematsu M, Hasegawa R, Imaida K, Tsuda H, Ito N (1983) Carcinogenesis. *4*:381
60. Peraino C, Fry RJM, Staffeldt E (1971) Cancer Res. *31*:1506
61. Hennings H, Shores R, Wenk ML, Spangler EF, Tarone R, Yuspa SH (1983) Nature, London *304*:67
62. Pitot HC, Goldsworthy T, Moran S, Sirica AE, Weeks J (1982) In: Hecker E (ed) Carcinogenesis: A Comprehensive Survey, Vol. 7. Raven Press, New York, p 85
63. Ghoshal AK, Ahluwahlia M, Farber E (1983) Am. J. Pathol. *113*:309
64. Pound AW, McGuire LJ (1978) Br. J. Cancer *37*:595
65. Roomi MW, Ho RK, Sarma DSR, Farber E (1985) Cancer Res. *45*:564
66. Ogawa K, Medline A, Farber E (1979) Lab Investig. *41*:22
67. Rotstein JR, Farber E (1986) Cancer Res.
68. Hayes MA, Roberts E, Dugan FP Cancer Res. Submitted for publication
69. Mori H, Furuya H, Williams GM (1983) J. Natl. Cancer Instit. *71*:849
70. Hayes MA, Lee G, Tatematsu M, Farber E (1987) Int. J. Cancer. In Press
71. Ward JM (1983) J. Natl. Cancer Instit. *71*:815
72. Ho RK, Mishkin S, Farber E (1982) Proc. Amer. Assoc. Cancer Res. *23*:52
73. Oesterle D, Deml E (1984) Carcinogenesis *5*:351
74. Hayes MA, Safe SH, Armstrong D, Cameron RG (1985) J. Natl. Cancer Instit. *74*:1037
75. Berry DL, Slaga TJ, DiGiovanni J, Juchau MR (1979) Ann. N.Y. Acad. Sci. *320*:405
76. Morales NM, Matthews HB (1979) Chem. Biol. Interact. *27*:99
77. Hargraves WA, Allen JR (1979) Res. Commun. Chem. Pathol. Pharmacol. *25*:543
78. Preston BD, Miller JA, Miller EC (1983) J. Biol. Chem. *258*:8304
79. Wong A, Basrur PK, Safe SH (1979) Res. Commun. Chem. Pathol. Pharmacol. *24*:543
80. Stadnicki S, Lin FSD, Allen JR (1979) Res. Commun. Chem. Pathol. Pharmacol. *24*:313
81. Hsia MT, Lin FS, Allen JR (1978) Res. Commun. Chem. Pathol. Pharmacol. *21*:485
82. Schoeny R (1982) Mutat. Res. *10*:45
83. Kimbrough RD, Linder RE, Gaines TB (1972) Arch. Environ. Health. *25*:354
84. Ito N, Nagasaki H, Makiura S, Arai M (1974) Gann. *65*:545
85. Nagasaki H, Tomii S, Mega T, Marugami M, Ito N (1972) Gann. *68*:805
86. Kimura NT, Baba T (1973) Gann. *64*:105
87. Schaeffer E, Greim H, Goessner W (1984) Toxicol. Appl. Pharmacol. *75*:278
88. Norback DH, Weltman RH (1985) Environ. Health Perspect. *60*:97
89. Morgan RW, Ward JM, Hartman PE (1981) Cancer Res. *41*:5052
90. Ward JM (1985) Environ. Health Perspect. *60*:89
91. Schulte-Hermann R, Timmerman-Troisiener I, Schuppler J (1983) Cancer Res. *43*:839
92. Alvares AP, Kappas A (1977) Clin. Pharmacol. Therap. *22*:809
93. Hollstein M, McCann J, Angelosanto FA, Nichols WW (1979) Mutat. Res. *65*:133
94. Deml E, Oesterle D, Wiebel FJ (1983) Cancer Lett. *19*:301
95. Anderson LM, Van Havere K, Budinger JM (1983) J. Natl. Cancer Instit. *71*:157
96. Shelton DW, Hendricks JD, Bailey GS (1984) Toxicol. Lett. *22*:27
97. Nishizumi M (1980) Gann. *71*:910
98. Makiura S, Aoe H, Sugihara S, Hirao K, Arai M, Ito N (1974) J. Natl. Cancer Instit. *53*:1253
99. Hendricks JD, Putnam TP, Bills DD, Sinhuber RO (1977) J. Natl. Cancer Instit. *59*:1545
100. Shelton DW, Hendricks JD, Coulombe RA, Bailey GS (1984) J. Toxicol. Environ. Health *13*:649
101. Bigger CAH, Tomaszewski JE, Dipple A, Lake RS (1980) Science. *209*:503
102. Hongslo J, Haug LT, Wirth PJ, Møller M, Dybing E, Thorgiersson SS (1983) Mutat. Res. *107*:239
103. Kohli KK, Mukhtar H, Bend JR, Albro PW, McKinney JD (1979) Biochem. Pharmacol. *28*:1444
104. Sharma RN, Cameron RG, Farber E, Griffen MJ, Joly JG, Murray RK (1979) Biochem. J. *183*:317
105. Ahotupa M (1981) Biochem. Pharmacol. *30*:1866
106. Parkinson A, Safe SH, Robertson LW, Thomas PE, Ryan DE, Reik LM, Levin W (1983) J. Biol. Chem. *58*:5967

107. Hayes MA, Roberts E, Roomi MW, Safe SH, Farber E, Cameron RG (1984) Toxicol. Appl. Pharmacol. *76*:118
108. Hayes MA, Roberts E, Safe SH, Farber E, Cameron RG (1986) J. Natl. Cancer Instit. *16*:683
109. Mitchell JR, Hughes H, Lauterburg BH, Smith CV (1982) Drug Metab. Rev. *13*:539
110. Farber JL, Gerson RJ (1984) Pharmacol. Rev. *36*:71S
111. Pereira MA, Herren SL, Britt AL, Khoury MM (1982) Cancer Lett. *15*:185
112. Pitot HC, Goldsworthy T, Moran S, Sirica AE, Weeks J (1982) Carcinogenesis. *7*:85
113. Sleight S (1985) Environ. Health Perspect *60*:35
114. Hirose M, Shirai T, Tsuda H, Fukushima S, Ogiso T, Ito N (1981) Carcinogenesis. *2*:1299
115. Collins WT, Capen CC, Kasza L, Carter C, Dailey RE (1977) Amer. J. Pathol. *89*:119
116. Collins WT, Capen CC (1980) Virchows Arch. B. Cell Path. *33*:213
117. Diwan BA, Palmer AE, Ohshima M, Rice JM (1985) J. Natl. Cancer Instit. *75*:1099
118. Hiasa Y, Kitahori Y, Oshima M, Fujita T, Yuasa T, Konishi N, Miyashiro A (1982) Carcinogenesis *3*:1187
119. Alvares AP, Kappas A (1977) J. Biol. Chem. *252*:6373
120. Alvares AP, Kappas A (1975) FEBS Lett. *50*:172
121. Wong TK, Everson RB, Hsu ST (1985) Lancet *1*:721
122. Ueng T-H, Alvares AP (1985) Toxicology *35*:83
123. Lake BG, Collins MA, Harris RA, Gangoli SD (1979) Xenobiotica. *9*:723
124. Allen JR (1975) Fed. Proc. *34*:1675
125. McConnell EE, Hass JR, Altman N, Moore JAA (1979) Lab. Anim. Sci. *29*:666
126. Schulte-Hermann R (1974) CRC Crit. Rev. Toxicol. *3*:97
127. Schulte-Hermann R, Parzefall W (1981) Cancer Res. *41*:4140
128. Ito N, Tsuda H, Hasegawa R, Imaida K (1983) Environ. Health Perspect. *50*:131
129. Reddy JK, Lalwai ND (1983) CRC Crit. Rev. Toxicol. *12*:11
130. Lans M, de Gerlache J, Taper HS, Preat V, Roberfroid M (1983) Carcinogenesis *4*:141
131. Columbano A, Ledda G, Rao PM, Rajalakshmi S, Sarma DSR (1984) Amer. J. Pathol. *116*:441
132. Williams GM (1981) Food Cosmet. Toxicol. *19*:577
133. McLean AEM, Marshall A (1971) Br. J. Exp. Pathol. *52*:322
134. Flaks A, Flaks B (1982) Carcinogenesis *3*:381
135. Hayes MA (1983) Proc. Amer. Assoc. Cancer Res. *24*:338
136. Hayes MA, Cameron RG, Roberts E, Safe S, Farber E (1984) Proc. Amer. Assoc. Cancer Res. *25*:505
137. Thorgiersson SS, Erickson LC, Smith CL, Glowinski IB (1983) Environ. Health Perspect. *49*:141
138. Haseman JK (1983) Fund. Appl. Toxicol. *3*:1
139. Roe FJC (1983) Nature. *303*:657
140. Clemmeson J, Fuglsang-Fredriksen V, Plum CM (1974) Lancet *1*:705
141. Clemmeson J, Hjalgrim-Jensen S (1978) Ecotoxicol. Environ. Safety. *1*:457
142. Tschirley FH (1986) Scientific American *254*:29
143. Roberts DW (1982) J. Amer. Med. Assoc. *247*:2142
144. Wolff MS, Anderson HA, Selikoff IT (1982) J. Amer. Med. Assoc. *247*:2112
145. Moseley CL, Geraci CL, Burg J (1982) Amer. J. Ind. Hygiene *43*:170
146. Mes J, Davies DJ, Turton D (1982) Bull. Environ. Contam. Toxicol. *28*:97
147. Unger M, Olsen J, Clausen J (1982) Environ. Res. *29*:371
148. Brown DP, Jones M (1981) Arch. Environ. Health *36*:120
149. Cammarano G, Crosignani P, Berrino F, Berra G (1984) Scand. J. Work. Environ. Health *10*:259
150. Couch JA, Harshbarger JC (1985) Environ. Carcinogenesis Res. *3*:63
151. Mix MC (1985) Mar. Environ. Res. (In press)
152. Malins DC, McCain BB, Brown DW, Chan S, Meyers MS, Landahl JT, Prohaska PG, Friedman AJ, Rhodes DG (1984) Environ. Sci. Technol. *18*:705
153. Maxim LD, Harrington L (1984) Regul. Toxicol. Pharmacol. *4*:192

Biotransformation of PCBs:
Metabolic Pathways and Mechanisms

I. G. Sipes [1] and R. G. Schnellmann [2]

Biotransformation of PCBs to hydroxylated metabolites by the microsomal cytochrome P-450 system is the critical event that determines the biological half-lives of these widespread environmental contaminants. The factors that determine the rate of biotransformation of PCBs are: the number of chlorines on the biphenyl nucleus, the position of these chlorines and the animal species. In general, as the number of chlorines increase and as the number of unsubstituted, adjacent carbon atoms decrease, the rate of cytochrome P-450 catalyzed hydroxylation of PCBs decreases. Evidence suggests that PCBs are hydroxylated via an arene-oxide intermediate, although other mechanisms may also be operative. The reactive arene oxides formed during biotransformation can bind covalently to tissue macromolecules and/or conjugate with glutathione. Derivatives of glutathione conjugates are major excretion products of PCBs as are glucuronides of the hydroxylated products. The species' differences in the rate of metabolism of PCBs can be related, at least in part, to differences in the basal levels of particular isozymes of cytochrome P-450 present in liver. The dog, which eliminates PCBs more rapidly than other species, possesses higher levels of a constitutive isozyme of cytochrome P-450 with activity towards the slowly metabolized, bioaccumulated, 2,2',4,4',5,5'-hexachlorobiphenyl. Future studies will determine the factors that regulate the activity of this particular isozyme as well as determine its activity toward other PCBs.

List of Abbreviations

245-HCB	2,2',4,4',5,5'-hexachlorobiphenyl
236-HCB	2,2',3,3',6,6'-hexachlorobiphenyl
246-HCB	2,2',4,4',6,6'-hexachlorobiphenyl
235-HCB	2,2',3,3',5,5'-hexachlorobiphenyl
2,5-TCB	2,2',4,4'-tetrachlorobiphenyl
4-DCB	4,4'-dichlorobiphenyl
P-450	cytochrome P-450

[1] Department of Pharmacology and Toxicology, College of Pharmacy, University of Arizona, Tuscon, AZ 85721, USA
[2] Department of Physiology and Pharmacology, College of Veterinary Medicine, University of Georgia, Athens, GA 30602, USA

1 Introduction

Biotransformation is the critical event that determines the biological fate of polychlorinated biphenyls (PCBs). In general, this class of compounds is poorly metabolized and thus elimination is slow. The slow rate of excretion can result in the bioaccumulation of PCBs resistant to metabolism. Bioaccumulation of PCBs has been associated with a number of toxicities. For example, the tumor promoting effects of PCBs appear to be caused by the parent compound. The virtually non-metabolized PCB, 2,2′4,4′,5,5′-hexachlorobiphenyl (245-HCB), promotes the development of hepatocarcinogenesis in rats while 2,2′,3,3′,6,6′-hexachlorobiphenyl (236-HCB), which is metabolized and excreted, is not a promoter of hepatocarcinogenesis (1, 2).

Just as the lack of metabolism of PCBs can result in toxicity, the process of metabolism has also been implicated in certain PCB-induced toxicities. Allen and Norback (3), Wyndham et al. (4) and others have suggested that the cytotoxic and mutagenic effects of PCBs result from the interaction of reactive intermediates (i.e. arene oxides) that are formed during metabolism with critical cellular processes. For example, the mutagenicity of 4-chlorobiphenyl is expressed only following its metabolism (4). *In vitro* studies using a Chinese hamster ovary (CHO) cell line demonstrated that 4-chlorobiphenyl was metabolized to hydroxylated products and that 4-chlorobiphenyl-equivalents bound covalently to DNA, RNA and protein (5). Furthermore, 4-chlorobiphenyl or a metabolite caused DNA damage in CHO cells as demonstrated by the induction of unscheduled DNA synthesis.

Stable metabolites of PCBs may also be responsible for toxicity. Bergman et al. (6) have reported the accumulation of methylsulfonyl metabolites of PCBs in the bronchi of mice. While the toxicological significance of this finding is not known, it has been suggested that these metabolites may be responsible for the persistent respiratory distress seen in the victims of PCB poisonings in Japan and the decreased lung vital capacity in workers exposed to PCBs (6–9).

The central role of metabolism in the biological half lives of PCBs is apparent from their pharmacokinetic profiles. In general, once PCBs are absorbed, they are initially distributed to liver and muscle (10). The relatively large amount of a dose stored in muscle is due to its large mass and high perfusion rate while storage in the liver results from its high perfusion rate and the affinity of PCBs for the liver. The redistribution of PCBs to the skin and adipose tissue is a result of a higher affinity of PCBs for these tissues (11). Ultimately, the predominant storage tissue for PCBs is adipose tissue (10). At equilibrium, any change in PCB concentration or change in volume of any tissue results in a corresponding change in all tissues. For example, if the concentration of a PCB in the liver is decreased by metabolism and excretion, then the concentration of that PCB in all tissues will decrease proportionally (10). This is illustrated by the data presented in Figure 1 which compared dog and monkey liver and adipose tissue concentrations of 245-HCB and its metabolites over time (12). Although 245-HCB does distribute to the adipose tissue of the dog, the storage in this tissue is not irreversible. As the liver continues to metabolize 245-HCB, the concentration of 245-HCB in the adipose tissue falls

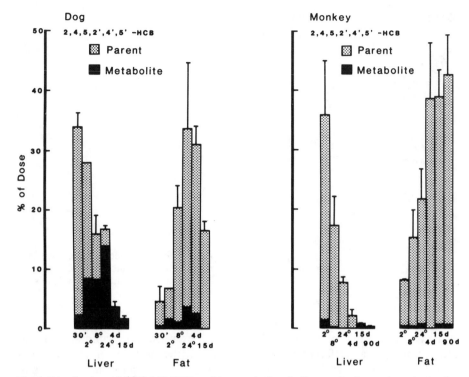

Fig. 1. Distribution of (^{14}C) 245-HCB and its metabolites in liver and adipose tissue in the dog and monkey. Taken from Sipes et al. (12) with permission

(day 15). In the monkey, however, under normal physiological conditions the storage of 245-HCB in adipose tissue is an irreversible event. The key difference between the dog and monkey is reflected by the concentration of metabolites present in the liver. Monkeys cannot readily metabolize 245-HCB, as seen by the lack of metabolites present in their livers, and therefore a major means of altering the equilibrium between tissue and blood pools is essentially absent.

Since metabolism plays such an important and central role in the toxicity and elimination of PCBs, the remaining portion of this chapter will focus on the *in vivo* and *in vitro* metabolism of PCBs. The key event in determining the rate of metabolism and elimination of PCBs is the biotransformation activity of hepatic cytochrome P-450(s) toward specific congeners. The *in vivo* metabolism of mono-, di-, tri-, tetra-, penta-, and hexachlorobiphenyls has been examined in a number of species (see 12–18). These studies showed that the rate of metabolism of these compounds depends on: 1) the number and placement of chlorines on the phenyl rings, and 2) the animal species.

In studies using rats, Matthews and Anderson found that the elimination half-lives of four PCBs with 1, 2, 5 or 6 chlorines increased as the number of chlorines increased (11). The decreased rate of elimination with increasing chlorination was directly related to the decreased rate of metabolism of the more highly chlorinated congeners. Studies in the dog have revealed the importance of chlorine position

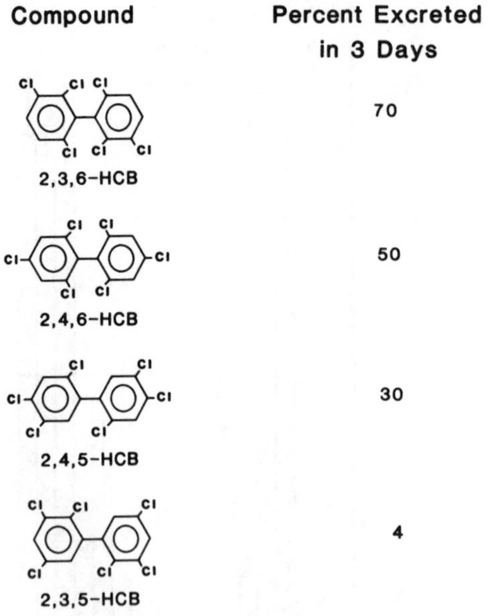

Compound	Percent Excreted in 3 Days
2,3,6–HCB	70
2,4,6–HCB	50
2,4,5–HCB	30
2,3,5–HCB	4

Fig. 2. The role of chlorine placement in the excretion of hexachlorobiphenyls by the dog following intravenous administration

in the biotransformation of PCBs (16). The dog excreted four symmetrical hexachlorobiphenyls at different rates depending on the position of the chlorines (Fig. 2). In this case, as the number of unsubstituted meta-positions or adjacent, unsubstituted carbon atoms increased, the percent of the dose excreted in three days increased. The rank order of excretion was 236-HCB > 2,2',4,4',6,6'-hexachlorobiphenyl (246-HCB) > 245-HCB > 2,2',3,3',5,5'-hexachlorobiphenyl (235-HCB). Thus, not only does the number of chlorines affect the rate of biotransformation but the position of the chlorines on the phenyl rings is also a critical factor.

Finally, the degree of biotransformation depends on the animal species. For example, the dog and rat eliminate 50 percent of a dose of 236-HCB as metabolites in one day while the monkey requires six days to eliminate the same amount (15, 18). For the 245-HCB congener, the dog eliminated 50 percent of a dose in eight days, while extrapolation of the elimination profiles from the monkey and rat showed that they were incapable of eliminating 50 percent of the administered dose during their remaining lifespans (12, 14, 15).

2 Mechanistic Studies

2.1 PCB Hydroxylation

The highly lipophilic PCBs lack functional groups and thus it is not surprising that they are initially hydroxylated by the hepatic cytochrome P-450 enzyme system (19). This was first demonstrated by showing that monochlorobiphenyl me-

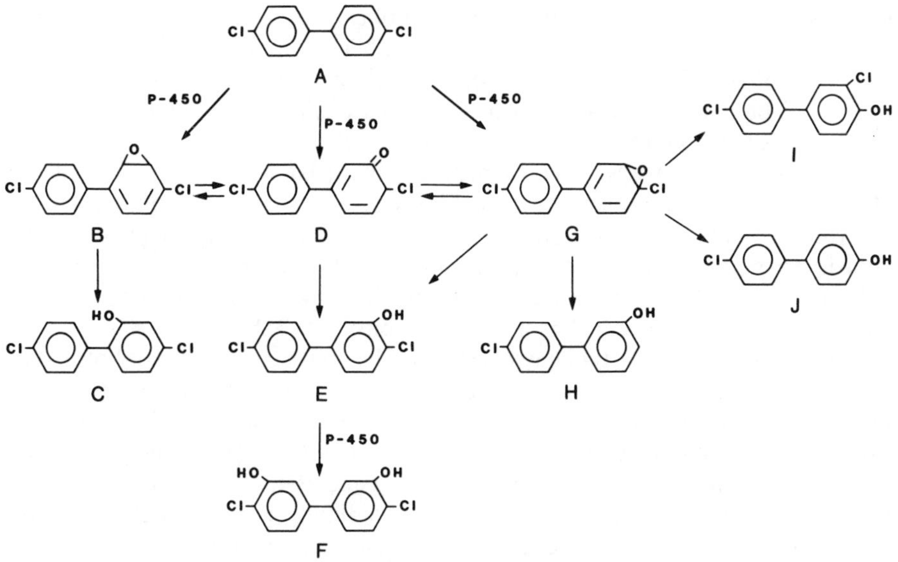

Fig. 3. Metabolites and possible pathways of 4,4′-dichlorobiphenyl (4-DCB) metabolism by human liver microsomes

tabolism is dependent on NADPH and oxygen and is inhibited by SKF-525A, metyrapone and carbon monoxide (20). As will be discussed later, studies with a reconstituted cytochrome P-450 system provided further support for the critical role of the mixed function oxygenases in PCB metabolism.

While PCBs are initially metabolized by the cytochrome P-450 system, little is known about the specific P-450 isozymes responsible for their metabolism. Kaminsky and colleagues (21, 22) reported that the major cytochrome P-450s induced in Sprague-Dawley rat livers by phenobarbital (PB-B) and beta-naphthoflavone (BNF-B) were capable of metabolizing a series of dichlorobiphenyls. The hepatic P-450 isozyme PB-B preferentially metabolized dichlorobiphenyls with noncoplanar conformations while the hepatic P-450 isozyme BNF-B preferentially metabolized coplanar dichlorobiphenyls. The conformational planarity of dichlorobiphenyls depends upon the presence or absence of ortho-chlorinated substituents. Dichlorobiphenyls with two ortho-chlorinated substituents are non-coplanar and thus were preferentially metabolized by isozyme P-450 PB-B. Dichlorobiphenyls with only one ortho-chlorinated substituent were metabolized by both P-450 PB-B and P-450 BNF-B at approximately equivalent rates. Another interesting finding of this work was the preferential hydroxylation of the chlorinated phenyl ring by P-450 BNF-B and the preferential hydroxylation of the non-chlorinated ring by P-450 PB-B.

Depending upon the degree of chlorination, the number and type of hydroxylated metabolites can vary in complexity. To illustrate the complex hydroxylation patterns that can occur with PCBs, the biotransformation of two model compounds [4,4′-dichlorobiphenyl (4-DCB) and 245-HCB] will be discussed. As demonstrated in Figure 3 the *in vitro* metabolism of 4-DCB (A) by human liver microsomes yields a variety of hydroxylated metabolites (23). The most abundant metabolite was 4,4'-dichloro-3-biphenylol (E), with lesser amounts of 3,4'-dichloro-4-biphenylol (I) 4'-chloro-4-biphenylol (J), 4'-chloro-3-biphenylol (H), and 4,4'-dichloro-2-biphenylol (C) being formed. A dihydroxylated metabolite, tentatively identified as 4,4'-dichloro-3,3'-biphenyldiol, was also found (F).

The possible pathways and intermediates involved in the formation of the above metabolites are also shown in Figure 3. The major metabolite, 4,4'-dichloro-3-biphenylol (E), is thought to form from the rearrangement of a keto-enol intermediate (D) or the 3,4-epoxide (G), whereas the metabolite 4,4'-dichloro-2-biphenylol (C) is thought to form by the rearrangement of a 2,3-epoxide (B). A NIH shift can occur during the metabolism of 4-DCB as seen by the formation of 3,4'-dichloro-4-biphenylol (I) (24). The presence of this metabolite supports the formation of an arene oxide as an intermediate in the metabolism of PCBs. These studies also showed that chlorines can be lost during PCB metabolism (4'-chloro-4-biphenylol, J; 4'-chloro-3-biphenylol, H). While the mechanism of chlorine loss has not been extensively investigated, it may have occurred during the rearrangement of the 3,4-epoxide (G). The dihydroxylated PCB metabolite (F) probably results from recycling of primary metabolites through the cytochrome P-450 system.

Similar hydroxylation patterns were seen when the highly chlorinated congener 245-HCB was incubated with rat liver microsomes (15). The major metabolite is 2,2',4,4',5,5'-hexachloro-3-biphenylol (E) with lesser amounts of 2,2',4,4',5,5'-hexachloro-6-biphenylol (C), 2,2',3,4',5,5'-hexachloro-4-biphenylol (G), 2,2',4',5,5'-pentachloro-3-biphenylol (I) and 2,2',4,5,5'-pentachloro-4-biphenylol (H) being formed (Fig. 5).

The pathways and intermediates of 245-HCB metabolism are similar to those of 4-DCB (Fig. 4). The major metabolite, 2,2',4,4',5,5'-hexachloro-3-biphenylol, is thought to form from the rearrangement of a keto-enol intermediate (D) or a 3,4-epoxide (F), whereas the ortho-hydroxylated metabolite (C) is thought to re-

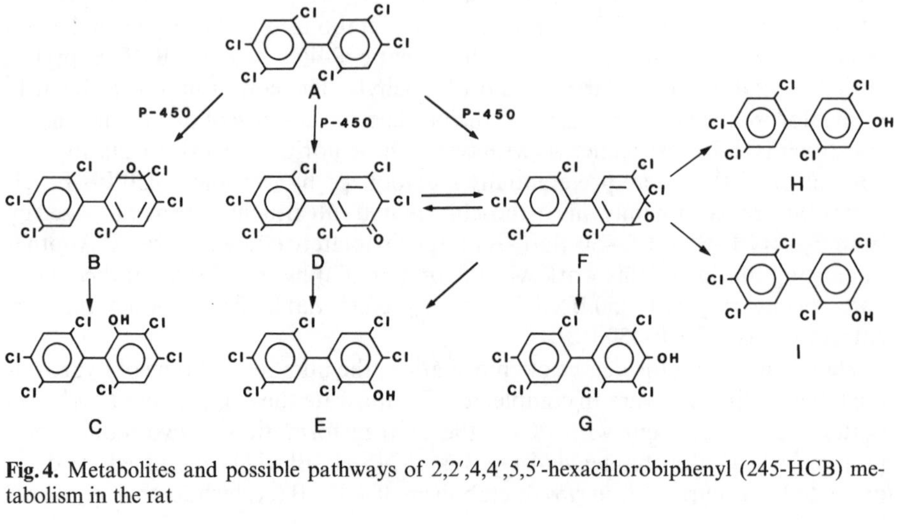

Fig. 4. Metabolites and possible pathways of 2,2',4,4',5,5'-hexachlorobiphenyl (245-HCB) metabolism in the rat

sult from the rearrangement of a 5,6-epoxide (B). A NIH shift can occur during the metabolism of this congener as seen by the formation of 2,2',3,4',5,5'-hexa-chloro-4-biphenylol (G) (24). The presence of this metabolite also supports the formation of an arene oxide intermediate in the metabolism of this PCB. The mechanism of dechlorination (H and I) is not known but may occur during the rearrangement of the 3,4-epoxide. Thus, the biotransformation pathways of 4-DCB and 245-HCB are similar and indicative that other PCBs have similar pathways of metabolism.

The detailed mechanism(s) of PCB hydroxylation are not known. Gardner et al. (25) provided the first evidence for the involvement of an arene oxide. Following administration of 2,2',5,5'-tetrachlorobiphenyl (2,5-TCB) to rabbits, they isolated from urine the metabolite, trans-3,4-dihydroxy-3,4-dihydro-2,5-TCB. This metabolite most likely forms from the reaction of a 3,4-epoxide of 2,5-TCB with water, either directly or with the aid of epoxide hydrolase. The phenolic metabolites formed from this compound may arise from the rearomatization of the dihydrodiol or keto-enol tautomer of the epoxide (13). Soon after this work was published, Safe, Hutzinger and co-workers examined the metabolism of 4-chlorobiphenyl in rabbits and frogs (13, 26). Using 4'-chloro-4-^{2}H-biphenyl as a substrate, rabbits and frogs were found to produce the same major metabolite, 4'-chloro-4-biphenylol, with 79 percent retention of the deuterium. These results are consistent with the initial formation of a 3,4-epoxide. Upon isomerization, the deuterium atom migrates from the site of final hydroxylation to a neighboring carbon atom. Daly et al. (24) have suggested that a shift of a deuterium atom or a halogen is evidence for the formation of an arene oxide intermediate. Further proof of an arene oxide intermediate in the metabolism of PCBs was provided by Forgue et al. (27) and Forgue and Allen (28). Incubation of 2,5-TCB with hepatic microsomes obtained from rats pretreated with phenobarbital resulted in formation of the 3,4-epoxide of 2,5-TCB, which they were able to isolate and identify. In addition, they obtained evidence of arene oxide formation in studies using hepatic microsomes obtained from rhesus monkeys pretreated with phenobarbital.

Data obtained from studies with human hepatic microsomes and 4-DCB as the substrate further support the involvement of an arene oxide in PCB metabolism. The 4-DCB metabolite, 3,4'-dichloro-4-biphenylol, is the product of a NIH shift from the intermediate 4,4'-dichloro-3,4-epoxy-biphenyl. The fact that 4-DCB equivalents bind covalently to microsomal protein and that this binding is reduced by glutathione further supports formation of a reactive intermediate (i.e., arene oxide) during the P-450 metabolism of this as well as other PCBs (23).

While the formation of an arene oxide as an intermediate in PCB metabolism has been established, evidence has been presented that suggests that PCBs and other aromatic halogenated hydrocarbons are also metabolized by non-arene oxide mechanisms. The most direct evidence for this pathway was first described for chlorobenzene. Selander and co-workers observed that chlorobenzene was metabolized to meta-chlorophenol in the rat and by rat liver microsomes (29, 30). However, this product was not the result of the rearrangement of either 2,3- or 3,4-chlorobenzene oxide. Preston et al. (31) have offered evidence that PCBs may also be metabolized by a non-arene oxide pathway. Using microsomes obtained from phenobarbital-treated rats, they demonstrated that 2,5-TCB was converted

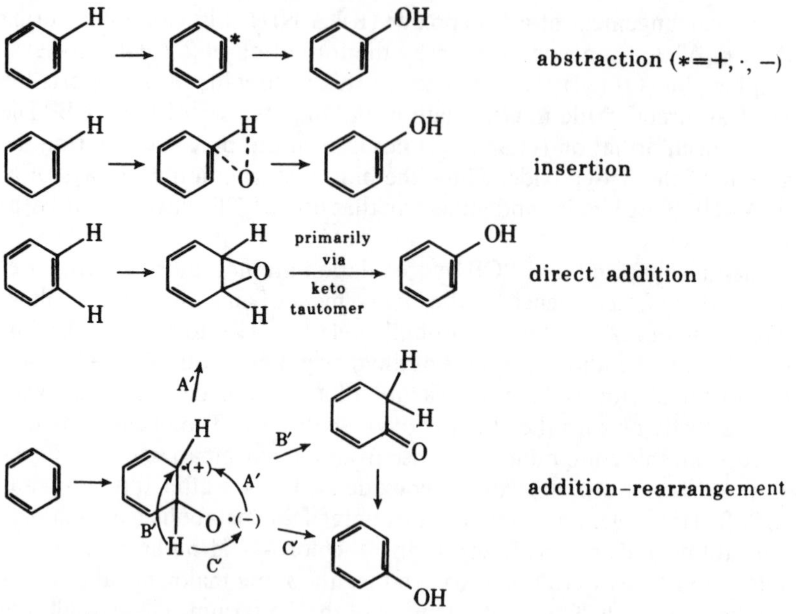

Fig. 5. Proposed mechanisms of aromatic hydroxylation. Taken from Tomaszewski et al. (1975) with permission

to 3-hydroxy-2,5-TCB with greater efficiency (eight-fold) than was the 3,4-oxide of 2,5-TCB converted to this same hydroxylated metabolite. These results suggest that the formation of 3-hydroxy-2,5-TCB did not occur exclusively through an arene oxide intermediate.

Tomaszewski et al. (32) proposed four mechanisms of aromatic hydroxylation: abstraction, direct addition, insertion and addition rearrangement (Fig. 5). Abstraction mechanisms of aromatic hydroxylation are usually excluded because of energetic reasons and because deuterium is retained in the hydroxylated products. The direct addition pathway proved popular to explain para-hydroxylation of biphenyl and chlorinated biphenyls since 1) an arene oxide and various dihydrodiols have been isolated from microsomal incubations of PCBs (25, 28), 2) a NIH shift with high deuterium retention was demonstrated (13, 26) and 3) no isotope effects were demonstrated (32, 33). The direct insertion pathway appeared to explain meta-hydroxylation because of the isotope effects seen with biphenyl, nitrobenzene and methyl phenyl sulfone (32, 33). The addition rearrangement pathway was initially excluded because of the energetic requirements needed to demonstrate an isotope effect for meta-hydroxylation. Thus, two mechanisms of aromatic hydroxylation were thought to be needed to explain the various products formed and that different P-450 isozymes might result in hydroxylation by different mechanisms.

However, Swinney et al. (34) recently demonstrated that meta-hydroxylation of biphenyl did not proceed via an abstraction or an insertion mechanism. They suggested that these data and data from other studies support the addition rearrangement pathway. This pathway is interesting because one mechanism of hydroxylation could explain all the products formed. From an evolutionary perspective, development of one mechanism of aromatic hydroxylation would be simpler than multiple mechanisms. One line of evidence that suggested that the addition rearrangement pathway was possible in P-450 reactions was shown by Miller and Guengerich (35). Using trichloroethylene (TCE), they showed that TCE oxide was not an obligatory first step in TCE metabolism. Furthermore, a major metabolite of TCE, 2,2,2-trichloroacetaldehyde, was not formed from TCE oxide. This work is important because it limited the possible role of the direct addition pathway in the metabolism of this haloalkene.

A similar mechanism may operate for aromatic hydroxylation. Indeed, Burka et al. (36) have proposed that halobenzenes initially form a P-450 perferryl oxide intermediate at the meta position. This tetrahedral intermediate can close to form either a 2,3- or 3,4-oxide, which can rearrange to form the ortho- or para-substituted phenol, respectively. If the direction of closure is determined by orientation at the active site of the P-450 enzyme, then it is reasonable to assume that one P-450 isozyme may yield the 3,4-epoxide, while another will yield the 2,3-epoxide. To obtain a meta-hydroxylated product, either a 1,2-hydride shift can occur or a hydrogen atom can be lost in the tetrahedral intermediate to yield a ketone that rearranges to form the meta-phenol.

Recently, Hanzlik et al. (37) reported that monosubstituted benzenes undergo hydroxylation by a combination of mechanisms. Using control rat liver microsomes and biphenyl, they found that 64 percent of ortho-hydroxylations, 100 percent of meta-hydroxylations and 94 percent of para-hydroxylations occurred via an indirect mechanism (not direct addition). Furthermore, their data suggested that a keto-enol intermediate is common to all indirect-type hydroxylations. While it has not been determined whether these mechanisms are applicable to all PCBs, they are not inconsistent with the known data.

Although the picture is far from complete, the addition-rearrangement theory is attractive for explaining the metabolism of PCBs. Differences in the sites of hydroxylation could be explained by differences in the active sites of different P-450s. Differences in the active sites could influence product formation at two different stages. The first stage is concerned with the initial placement of the substrate. Depending on how the substrate is oriented, different carbon atoms (ortho-, meta-, or para-) may be attacked, initially. The second stage at which the active site may influence product formation is when the tetrahedral intermediate rearranges. Depending on the *milieu interieur* of the active site, the intermediate may rearrange to form an ortho,meta-epoxide; meta,para-epoxide or undergo a shift to form the meta-substituted phenol. Thus, different P-450 isozymes involved in aromatic hydroxylation may all activate the compound by a similar mechanism but differ by activating different carbons or influencing rearrangement products.

While the exact intermediates involved in PCB metabolism remain to be determined, the reactivity of these intermediates has been verified. Administration of 236-HCB to mice results in covalent binding of 236-HCB equivalents to RNA, DNA and protein (40). In contrast, 245-HCB, a virtually unmetabolized PCB congener, was found not to covalently bind to DNA and bound at least an order of magnitude less than 236-HCB to RNA and protein.

In *in vitro* microsomal systems, PCBs have been shown to bind to microsomal proteins by a cytochrome P-450 dependent mechanism which is thought to be epoxide-mediated (23, 38, 41–43). Induction of the cytochrome P-450 enzyme system stimulates the amount of covalent binding (41, 42). For example, covalent binding of 2,2′,4,4′-tetrachlorobiphenyl to protein in a reconstituted system is primarily mediated by the cytochrome P-450 induced by phenobarbital (P-450$_{PB}$) whereas covalent binding of 3,3′,4,4′-tetrachlorobiphenyl is mediated by 3-methylcholanthrene induced cytochrome P-450s (P-448$_{MC}$) (44). Analysis of the target proteins alkylated by the PCBs in the reconstituted system showed that several proteins were selectively alkylated, including cytochrome P-450 reductase and cytochrome P-448$_{MC}$. When a cytosolic fraction was added to the incubation, many cytosolic proteins were also targets for the reactive intermediates. Interestingly, when the covalent binding of 3,3′,4,4′-tetrachlorobiphenyl (mediated by a reconstituted P-448 system) was examined in the presence of amino acids or peptides with reduced sulfhydryl groups, or proteins with and without reduced sulfhydryl groups, ^{14}C-equivalents of tetrachlorobiphenyl selectively bound to the agents with the free-sulfhydryl groups. These results show that multiple P-450s may activate PCBs and that the targets of these reactive species are not random.

2.2 Species Variation in Rates of Hydroxylation

Differences in the rates of PCB metabolism and excretion among the species are most likely explained by differences in the basal levels of the different P-450 isozymes responsible for PCB metabolism, rather than by differences in reaction mechanisms. How the substrate is oriented at the active site will determine the regioselectivity of hydroxylation. Factors that control the nature of the active site may also influence the rearrangement of the tetrahedral intermediate. It is interesting that for the dog, the ratio of covalently bound to total metabolites is greater for 236-HCB than for 4-DCB (38). Perhaps a greater proportion of the 236-HCB-P-450-perferryl oxide complex rearranges to the 4,5-oxide, while for 4-DCB, a greater proportion rearranges to the ketone intermediate. It is also interesting that the rate of elimination by the dog of 235-HCB, a congener with no unsubstituted meta positions, was slower than that of the other hexachlorobiphenyls. Substitution of these positions may hinder formation of the meta-tetrahedral intermediate.

In support of the hypothesis that species differences in PCB metabolism are the result of different basal levels of P-450 isozymes, it was found that dog hepatic microsomes metabolize 245-HCB at a much greater rate than rat hepatic microsomes (Table 1). Pretreatment of dogs or rats with phenobarbital, increases the *in vitro* hepatic microsomal metabolism of 245-HCB in both species (Table 1). An isozyme of P-450 was isolated from the hepatic microsomes obtained from phenobarbital pretreated dogs that possessed high activity towards 245-HCB in a reconstituted system. Antibodies raised towards this isozyme of P-450 inhibited the metabolism of 245-HCB by hepatic microsomes obtained from control as well as

Table 1. Metabolism of 245-HCB by dog and rat hepatic microsomes

Enzymes source	Dog pmoles/mg protein in 20 min	Rat pmoles/mg protein in 20 min
Hepatic microsomes (Control animals)	460 ± 90	40 ± 10
Hepatic microsomes (Phenobarbital-induced animals)	2200 ± 570	80 ± 10

Results are expressed as the formation of total aqueous extractable metabolites per mg microsomal protein per 20 min, and represent the mean ± SD of the values obtained from control or phenobarbital-induced male beagle dogs (n=3) or Sprague-Dawley rats (n=4). Substrate concentration was 40 µM. Taken from Duignan et al. (39) with permission

phenobarbital-treated dogs (39). Thus, it appears that dog liver contains a higher basal concentration of a cytochrome P-450 isozyme which biotransforms 245-HCB, and probably other PCBs, to hydroxylated products. The higher concentration of this P-450 isozyme in dog liver explains why the dog can biotransform and eliminate PCBs more rapidly, and to a greater extent, than other species.

3 Conjugation Reactions

Conjugation of oxidative metabolites (phase II metabolism) is an important pathway of PCB metabolism but has received only limited attention. O-glucuronides of hydroxylated PCBs have been identified in the excreta of the rabbit, monkey and rat (9, 45, 46). PCB-glucuronides have been found also to form in *in vitro* systems. Human liver microsomes, in the presence of UDPGA, can glucuronidate 4,4'-dichloro-3-biphenylol, the major metabolite of 4-DCB (23). Methylation of hydroxylated PCBs is another identified conjugation process involved in PCB metabolism. The rat metabolizes 2,4,6-trichlorobiphenyl to 2,4,6-trichloro-3',4'-biphenyldiol which is then methylated at one of the hydroxyl groups (47). Following the introduction of the second hydroxyl group, the resulting dihydroxy compound is actually a para-substituted catechol. These catechols may show some affinity for the enzyme, catechol-O-methyltransferase, which methylates the neurotransmitters, norepinephrine and epinephrine.

The mercapturic acid pathway has also been shown to play a role in PCB metabolism. Bakke et al. (9) have proposed that an arene oxide of 2,4',5-trichlorobiphenyl reacts with glutathione (GSH) to form a GSH-conjugate (Fig. 6). This conjugate is then metabolized by gamma-glutamyl transpeptidase and dipeptidase to yield the cysteine-adduct. The cysteine-adduct can then be metabolized by intestinal bacteria and tissue C-S lyases to produce 2,4',5-trichlorobiphenylthiol. The thiol-adduct is converted to the methylthio-derivative by transmethylation from S-adenosylmethionine and the action of thio-S-methyltransferase.

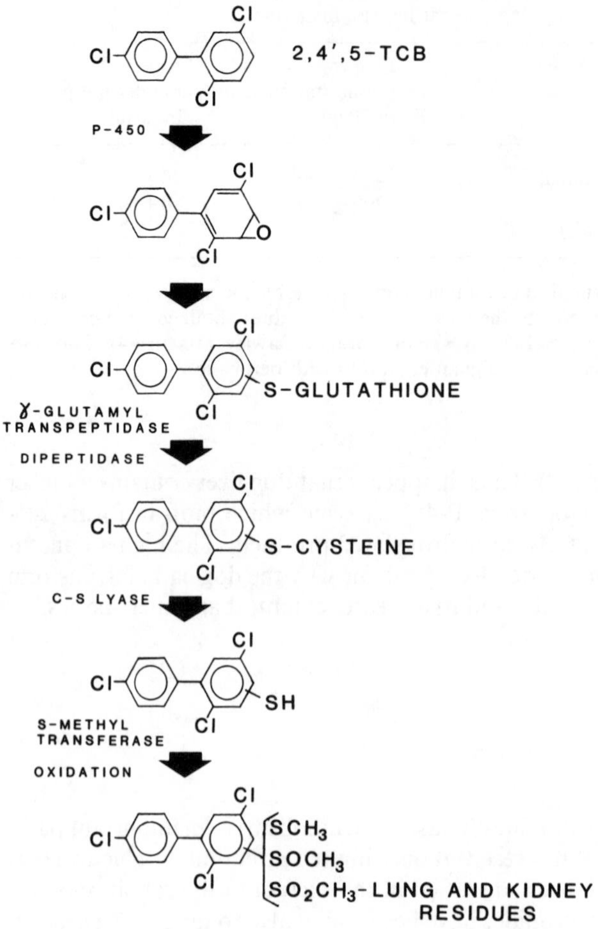

Fig. 6. The formation of mercapturic acid metabolites of PCBs. Taken from Bakke et al. (9) with permission

The methylthio-metabolite is thought to undergo oxidation by the mixed function oxygenase, first through a methylsulfinyl-metabolite and ultimately to a methyl-sulfonyl-metabolite (48). The methylthio-, methylsulfinyl- and methylsulfonyl metabolites of 2,4',5-trichlorobiphenyl isolated from the feces of treated rats accounted for 22 percent of the administered dose (9). Thus, the mercapturic acid pathway represents a significant route of PCB metabolism and is closely coupled with phase I metabolism.

4 Acknowledgements

This research supported by Grants ES-82130 and ES-07091 from the National Institutes of Health and a grant from Hoffmann-LaRoche, Inc. The authors thank Leslie Auerbach for her assistance with manuscript preparation.

5 References

1. Weltman RH, Norback DH (1979) Toxicol. Appl. Pharmacol. *48*:A181
2. Weltman RH, Norback DH (1983) Toxicologist *3*:401
3. Allen JR, Norback DH (1977) In: Hiatt HH, Watson JD, Winsten JA (eds) Incidence of Cancer in Humans. Cold Spring Harbor Laboratory, Cold Spring Harbor, New York, p 173
4. Wyndham C, Devenish J, Safe S (1976) Res. Commun. Chem. Pathol. Pharmacol. *15*:563
5. Wong A, Basrur PK, Safe S (1979) Res. Commun. Chem. Pathol. Pharmacol. *24*:543
6. Bergman A, Brandt I, Jansson B (1979) Toxicol. Appl. Pharmacol. *48*:213
7. Shigematsu N, Ishimura S, Saito R, Ikeda T, Matsuba K, Sugiyama K, Masuda Y (1978) Environ. Res. *16*:72
8. Warshaw R, Fischbein A, Thornton J, Miller A, Selikoff IJ (1979) Ann N.Y. Acad. Sci. *320*:277
9. Bakke JE, Bergman AL, Larsen GL (1982) Science *217*:645
10. Matthews HB, Dedrick RL (1984) Ann. Rev. Pharmacol. Toxicol. *24*:85
11. Matthews HB, Anderson MW (1975) Drug Metab. Dispos. *3*:371
12. Sipes IG, Slocumb ML, Perry DF, Carter DE (1982) Toxicol. Appl. Pharmacol. *65*:264
13. Sundstrom G, Hutzinger O, Safe S (1976) Chemosphere *5*:267
14. Lutz RJ, Dedrick RL, Matthews HB, Eling TE, Anderson MW (1977) Drug Metab. Dispos. *5*:386
15. Kato S, McKinney JD, Matthews HB (1980) Toxicol. Appl. Pharmacol. *53*:386
16. Sipes IG, McKelvie DH, Collins R (1979) Toxicol. Appl. Pharmacol. *48*:A155
17. Sipes IG, Slocumb ML, Perry DF, Carter DE (1980) Toxicol. Appl. Pharmacol. *55*:554
18. Sipes IG, Slocumb ML, Chen H-SG, Carter DE (1982) Toxicol. Appl. Pharmacol. *62*:317
19. Grant DL, Phillips WEJ, Villeneuve DC (1971) Bull. Environ. Contam. Toxicol. *6*:102
20. Kennedy MW, Carpenter NK, Dymerski PP, Adams SM, Kaminsky LS (1980) Biochem. Pharmacol. *29*:727
21. Kennedy MW, Carpenter NK, Dymerski PP, Kaminsky LS (1981) Biochem. Pharmacol. *30*:577
22. Kaminsky LS, Kennedy MW, Adams SM, Guengerich FP (1981) Biochem. *20*:7379
23. Schnellmann RG, Volp RF, Putnam CW, Sipes IG (1984) Biochem. Pharmacol. *33*:3503
24. Daly JW, Jerina DM, Witkop B (1972) Experientia *28*:1129
25. Gardner AM, Chen JR, Roach JAG, Ragelis EP (1973) Biochem. Biophys. Res. Commun. *55*:1377
26. Safe S, Hutzinger O, Jones D (1975) J. Agr. Food Chem. *23*:851
27. Forgue ST, Preston BD, Hargraves WA, Reich IL, Allen JR (1979) Biochem. Biophys. Res. Commun. *91*:475
28. Forgue ST, Allen JR (1982) Chem.-Biol. Interact. *40*:233
29. Selander HG, Jerina DM, Piccolo DE, Berchtold GA (1975) J. Am. Chem. Soc. *97*:4428
30. Selander HG, Jerina DM, Daly JW (1975) Arch. Biochem. Biophys. *168*:309
31. Preston BD, Miller JA, Miller EC (1983) J. Biol. Chem. *258*:8304
32. Tomaszewski JE, Jerina DM, Daly JW (1975) Biochem. *14*:2024
33. Billings RE, McMahon RE (1978) Mol. Pharmacol. *14*:145
34. Swinney DC, Howald WN, Trager WF (1984) Biochem. Biophys. Res. Commun. *118*:867

35. Miller RE, Guengerich FP (1982) Biochem. *21*:1090
36. Burka LT, Plucinski TM, Macdonald TL (1983) Proc. Natl. Acad. Sci. *80*:6680
37. Hanzlik RP, Hogberg K, Judson CM (1984) Biochem. *23*:3048
38. Schnellmann RG, Sipes IG (1984) Toxicologist *4*:95
39. Duignan DB, Sipes IG, Leonard TB, Halpert JR (Submitted for publication)
40. Morales NM, Matthews HB (1979) Chem.-Biol. Interact. *27*:99
41. Hesse S, Metzger M, Wolff T (1978) Chem. Biol. Interact. *20*:355
42. Shimada T, Sato R (1980) Toxicol. Appl. Pharmacol. *55*:490
43. Schnellmann RG, Putnam CW, Sipes IG (1983) Biochem. Pharmacol. *32*:3233
44. Shimada T, Sawabe Y (1983) Toxicol. Appl. Pharmacol. *70*:486
45. Block WD, Cornish HH (1959) J. Biol. Chem. *234*:3301
46. Norback DH, Mack E, Reddy G, Britt J, Hsia MT (1981) Res. Commun. Chem. Pathol. Pharmacol. *32*:71
47. Shimada T, Sato R (1980) Toxicol. Appl. Pharmacol. *55*:490
48. Mio T, Sumino K (1985) Env. Hlth. Perspect. *59*:129

Physiologic Pharmacokinetic Modeling
of Polychlorinated Biphenyls

R. J. Lutz [1] and R. L. Dedrick [1]

A flow-limited physiologic pharmacokinetic model is presented for the analysis of the distribution and disposition of some polychlorinated biphenyl congeners in several animal species. Analysis of the pharmacokinetic parameters in the model gives insight into congener-to-congener and species-to-species comparisons. For example, the partition coefficient for parent compound of all congeners is greatest in the fat tissue. Clearance of PCBs occurs predominantly as metabolites into the urine and feces. Rates of metabolism vary considerably among congeners in all species studied, and there is no apparent scaling factor, such as body weight or surface area, for predicting rates of metabolism from species to species. Physiologic pharmacokinetic models are useful didactic tools for assessing models of distribution of PCBs.

1 Introduction

Of the 1.4 billion pounds of polychlorinated biphenyls (PCBs) that have been produced in the United States, it is estimated that 750 million pounds are still in use in some form, mostly in electrical transformers and capacitors (1). PCBs have produced adverse toxic reactions in experimental animals (2, 3) and man (4–6). The persistence of such large quantities of potentially toxic PCBs and the ever-increasing distribution of other anthropogenic chemicals into our environment require more sophisticated methods for estimating body burdens in animals and in man in order to make rational assessments of the toxicological risks posed to the population. Pharmacokinetic analysis is a logical first step in approaching this problem since it involves the study of the time course of the uptake, distribution,

[1] Biomedical Engineering and Instrumentation Branch, Division of Research Services, National Institutes of Health, Bethesda, MD 20892, USA

and disposition of chemicals in living systems. It gives the time-dependent concentration of a substance in various organs or tissues and therefore provides information about potential toxicity after exposure by some route. Mathematical modeling can be applied to pharmacokinetic analysis as an attempt to organize and quantitate pharmacokinetic data. Recently, physiologic models have gained popularity as a method for pharmacokinetic analysis (7, 8). These models apply available physiological, anatomical, and biochemical information to predict and to simulate chemical disposition *in vivo*. This paper focuses on the applications of a physiologic pharmacokinetic model used to analyze the distribution of polychlorinated biphenyl in several animal species. No attempt is made to review the plethora of literature that exits on experimental pharmacokinetic data presented without model analysis. The following sections present a brief description of physiologic modeling principles and cite specific examples of model application to the pharmacokinetic data of several PCB congeners in mouse, rat, dog, and monkey. The discussion includes ways that pharmacokinetic models can be used to predict PCB distribution in humans by extrapolation of data from other animal species. More detailed descriptions of this type of model analysis are available in the literature (9–11).

2 Physiologic Models

The basic unit of construction of a lumped physiologic model is a compartment such as illustrated in Fig. 1. The compartment represents a region of the body such as a discrete organ like the kidney or liver, or a widely distributed anatomical region such as muscle or fat. A single compartment can be subdivided into plasma (blood), interstitial, and intracellular space. These are assumed to be spatially uni-

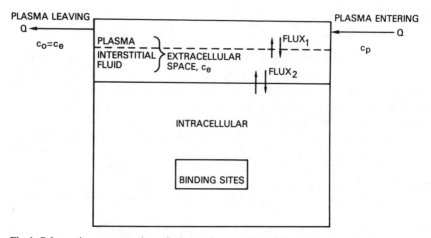

Fig. 1. Schematic representation of a lumped compartment in a pharmacokinetic model

form in concentration; however, the concentrations generally change with time. Transport of chemical occurs by the inflow (arterial) and outflow (venous) of blood through the compartment, by transcapillary transport (FLUX1), and by transport across cell membranes (FLUX2). For many exogenous chemicals the transcapillary and membrane transport rates are relatively rapid compared to the net input by blood flow. This introduces the simplifying concept of blood-flow-limited uptake which appears to be a valid assumption for most tissues used in the PCB model. The skin (12) is an important exception. Therefore, the analysis described in this chapter is for a flow-limited model of PCB. Membrane limited models are required for those chemicals that cross capillaries and cell membranes slowly with respect to blood flow rates. With rapid transcellular transport, equilibrium partitioning between blood and tissue can be assumed. Though this partitioning can include both nonspecific and specific saturable binding which may be nonlinear, the single dose of PCB used in our own pharmacokinetic studies warranted only the use of a constant partitioning coefficient. A more extensive description of these concepts is given by Lutz et al. (11).

Numerous compartments are arranged according to an anatomical flow scheme in order to simulate any animal species. Mammals show remarkable geometric similarity. A compartment is included in the model if it represents a region of substantial chemical uptake, if it is involved in a clearance process, such as excretion or metabolism, or if it is a site of special interest due to its toxic response to the chemical. Figure 2 presents the structure of the physiologic model that has been applied to the study of PCBs. It shows the relevant compartments, blood flows and excretory mechanisms. Each tissue is represented by flow-limited lumped compartments wherein the dashed lines symbolize rapid, equilibrium partitioning between blood and tissue space. The mathematical model is a set of mass balance equations on each compartment for parent PCB and its metabolites. So-

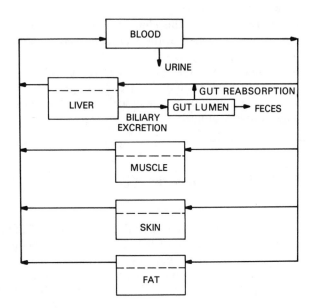

Fig. 2. Flow diagram for pharmacokinetic model of chlorinated biphenyls

lution of these equations provides the concentrations of PCB and its metabolites in each compartment as a function of time. For example, the material balance for the metabolite on the liver compartment yields the equation:

$$VL*dCL'/dt = QL[CB' - CL'/RL'] + kG*VG*CG' - kB*CL'/RL' + km*CL/RL$$

where

VL = volume of the liver
QL = blood flow rate to the liver
kG = gut readsorption rate constant
VG = volume of the gut lumen
kB = biliary clearance
km = metabolism rate constant
t = time
RL, RL' = equilibrium partitioning coefficient between blood and liver for parent amd metabolite, respectively
CB', CL', CG' = metabolite concentration in the blood, liver, and gut lumen, respectively
CL = parent concentration in the liver.

It is apparent that a substantial number of parameters are required in order to utilize this type of model for predicting chemical distribution. The parameters can be listed under four major categorical headings: (1) anatomic, such as tissue size and organ volumes; (2) thermodynamic, such as equilibrium partitioning or binding coefficients; (3) physiologic, such as clearances, metabolism rate constants, and blood flow rates; and (4) transport, such as diffusion coefficients and cell membrane permeabilities. For the case of the flow-limited PCB model, transport parameters are not relevant since the membrane permeabilities are considered to be large relative to blood flow rates. Many of the parameters necessary for the model are available from the literature, especially anatomical parameters and blood flow rates. Others can be deduced or approximated from preliminary experiments. A more detailed description of parameter estimation techniques is found in Lutz et al. (11).

The remainder of this chapter is devoted to a presentation of the results of physiologic modeling of the pharmacokinetic studies of several PCB congeners in various animal species. The model serves as a good didactic tool to help explain or rationalize experimental findings as they relate to interspecies correlations for a given congener or to differences in congener-to-congener distribution within a species. Examination of the model parameters puts these comparisons on more quantitative footing.

3 PCB Pharmacokinetics

This section presents a qualitative description of the pharmacokinetic behavior of PCBs. Experimental distribution data from the mouse, rat, dog, and monkey were used in the development of the model and for obtaining the parameters for these several species (13–18). The experiments were carried out using intravenous bolus injections of chemical. This procedure obviates the complications imposed by first having to investigate exposure and uptake behavior by non-parenteral routes such as oral ingestion, inhalation, or dermal absorption. Properly derived model parameters should have universal utility and be applicable to simulations of any route of administration. This concept highlights one of the advantages of physiologic pharmacokinetic models.

The PCB model was constructed initially to describe distribution data for 4-monochlorobiphenyl (4-MCB), 44'-dichlorobiphenyl (DCB), 22'455'-pentachlorobiphenyl (5CB) and 22'44'55'-hexachlorobiphenyl (HCB) in the rat. The same model was subsequently applied to these PCBs in the mouse (13). There are 209 possible congeners for PCB, making it essentially impossible to study each one independently. The compounds studied were chosen in order to demonstrate the effects of varying degrees of chlorination and the position of the chlorines on the biphenyl rings on the distribution and metabolism. For the subsequent dog and monkey studies, 44'-DCB, 22'44'55'-HCB and 22'33'66'-HCB were used. The latter two were selected since they have the same number of chlorines, except that the 22'33'66'-HCB has two adjacent unsubstituted carbon atoms on the phenyl rings. The presence of adjacent positions available for enzymatic attack has been shown to enhance the rate of metabolism (19).

After a PCB has been injected into an animal, there is a rapid redistribution phase in which the chemical passes from the blood into the various tissues. For PCBs, this process appears to be blood-flow limited for each tissue with the exception of skin (12). The redistribution kinetics of uptake in each tissue depend on the tissue blood perfusion rate per unit volume of PCB space. PCB space can be thought of as the anatomical tissue volume times the tissue-to-blood partition coefficient. As the diagram in Figure 2 shows, the tissue compartments are connected in parallel via the blood supply so that the whole-body kinetics depend on the complex interaction of individual tissues with blood.

Metabolism of parent PCB occurs in the liver by hepatic mixed function oxidases. Because of its high blood flow, the liver receives a large fraction of the parent PCB early and converts it to metabolite at rates depending on the particular PCB congener. The metabolites can be conjugated primarily to glucuronides which are excreted via the bile into the gut lumen. Some of the metabolite formed in the liver enters the hepatic blood which returns to the blood pool and redistributes to other tissues. Some fraction of the circulating metabolite is cleared by the kidney. More detailed quantitative descriptions of metabolic and clearance parameters are given in the later sections.

Consider two specific examples of PCB distribution in the rat, one the rapidly metabolized 4-monochlorobiphenyl (4-MCB), and the other the very slowly metabolized 22'44'55'-hexachlorobiphenyl (245-HCB). Figure 3 shows the concentration of 4-MCB parent and its metabolite (difference between total and parent) in the blood of rats after an iv injection. The points in these figures represent the experimental data, and the lines represent the model simulations. Due to its rapid metabolism in the liver and uptake in tissues, the parent PCB falls to less than 0.1% of peak concentration in 10 hours. The metabolite that is formed circulates in blood while redistributing to tissues as it is being cleared by the liver and kidney. The concentrations of parent and metabolite in various tissues of the rat are shown in Figure 4. Note that the muscle and liver have a concentration pattern

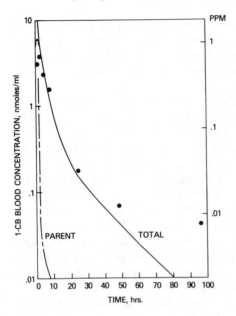

Fig. 3. 4-MCB blood concentration as a function of time after a single iv dose of 0.6 mg/kg in the rat. Points represent experimental data for total 1-CB. Simulations are given for total equivalents (——) and parent 1-CB (–––). Beyond 10 hr, total concentration in blood is composed almost entirely of metabolite. Reference (12)

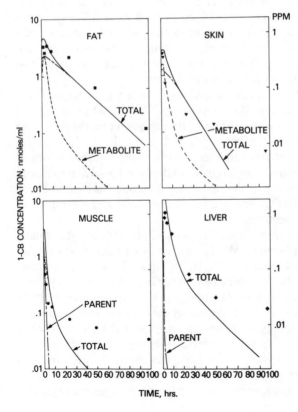

Fig. 4. 4-MCB tissue concentration as a function of time after a single iv dose of 0.6 mg/kg in the rat. Points represent experimental data for total 1-CB in the tissues. In each figure, total equivalent concentration (——), metabolite concentration (–––), and parent concentration (–––) are shown. Reference (12)

similar to blood with high metabolite-to-tissue ratios. However, even though blood concentration of parent is low after 10 hours, the fat and skin have higher parent concentrations and relatively low metabolite concentrations. This distribution reflects the high degree of partitioning of the lipophilic parent PCB into fat and skin which sequester parent PCB during the early redistribution phase before parent blood levels have been diminished by metabolism. The concentration of PCB between 10 and 100 hours in Figure 4 for fat and skin is much greater than that in blood. The slow removal from the fat space is controlled by the low blood perfusion through the fat tissue and the high fat-to-blood equilibrium distribution ratio. The relative concentrations of parent and metabolite are tissue dependent as a result of a combination of rapid metabolism, different tissue perfusion rates, and widely varying partitioning coefficients. Concentrations of parent and metabolite in blood alone do not, in general, reflect the same relative concentrations of parent and metabolite of all tissues.

Compared to the 4-MCB, the persistence of 245-HCB is much longer in all species studied. Except for the dog, metabolism of this HCB in the liver is extremely slow, which limits the rate of elimination. Figure 5 shows the blood concentration

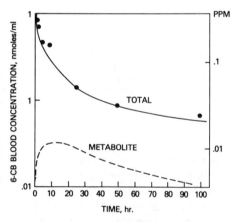

Fig. 5. 245-HCB blood concentration as a function of time after single iv dose of 0.6 mg/kg in the rat. Points represent experimental data for total 6-CB equivalents. Simulations are given for total (——) and metabolite (– – –). Reference (12)

Fig. 6. 245-HCB tissue concentration as a function of time for 96 hr after a single iv dose of 0.6 mg/kg in the rat. Points represent experimental data. Simulations of total 6-CB equivalents are shown by the solid lines. Very little metabolite appeared in any tissue. Reference (12)

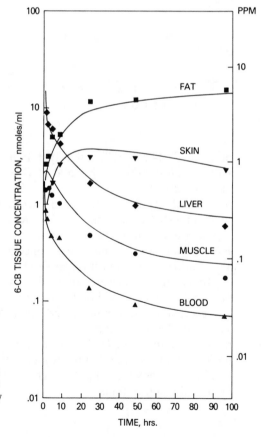

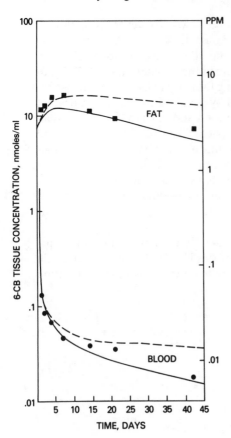

Fig. 7. 245-HCB concentration in fat and blood as a function of time for 42 days after a single iv dose of 0.6 mg/kg in the rat. The dashed lines represent simulations with a constant fat volume. The solid lines represent simulations with an increasing fat volume as described in (12)

of parent 22′44′55′-HCB and its metabolite in the rat. The metabolite levels represent a small fraction of the total PCB concentration. The distribution of slowly metabolized PCBs can be ascertained primarily from the pharmacokinetics of parent PCB. Tissue-to-blood concentration ratios are much higher for 245-HCB than for 4-MCB, and the metabolite concentration is neglible in all tissues. Figure 6 shows the concentration history of 245-HCB in the rat up to 96 hours. Even for this small animal, peak concentrations in the fat are not achieved until a week after a bolus injection, as shown by the data of Figure 7.

Model parameters were obtained for several PCB congeners in the mouse, rat, dog, and monkey, based on tissue concentration data for parent and metabolite and excretion data for urine and feces. Anatomical parameters (compartment sizes and blood flow rates) are listed in Table 1. Metabolism and clearance parameters are listed in Tables 2 and 3. An examination of these parameters gives useful insight into the dominant mechanisms controlling PCB distribution and into interspecies variations. The significance of the model parameters is discussed in the following sections.

Table 1. Volumes and flow rates in several tissues of four species. Values were used in physiologic pharmacokinetic model

	Volumes, ml				Blood flow rates, ml/min			
	Mouse[a] 30 g	Rat[b] 250 g	Monkey[a] 5 kg	Dog[c] 12 kg	Mouse 30 g	Rat 250 g	Monkey 5 kg	Dog 12 kg
Blood	2.89	22.5	300	1000	–	–	–	–
Muscle	17.1	125	2068	5530	1.42	7.5	103	275
Liver	2.24	10	118	480	3.10	16	125	342
Skin	5.51	40	470	1680	0.12	0.5	2.7	11.7
Fat	3.72	17.5	389	777	0.10	0.4	10.7	17.9

[a] Data from reference (13)
[b] Data from reference (12)
[c] Data from reference (30)

Table 2. Metabolism rate constant from physiologic model, k_m. Middle set of numbers is per unit animal body weight, ml/min/kg. Bottom set is $ml/min/kg^{.7}$, since body surface area is approximately proportional to body weight to the 0.7 power. References (12, 13, 30)

	4-MCB	44′-DCB	22′455′-PCB	245-HCB	236-HCB
k_m, ml/min					
Mouse (38 g)	2.4	0.37	0.095	0.01	–
Rat (250 g)	10.0	2.0	0.39	0.045	5.0
Monkey (5 kg)	–	7.0	–	0.67	15.0
Dog (12 kg)	–	470	–	16.0	183
k_m, ml/min/kg					
Mouse	68.5	9.7	2.5	0.25	–
Rat	40.0	8.0	1.56	0.18	20
Monkey	–	1.4	–	0.13	3.0
Dog	–	39	–	1.33	15.2
k_m, ml/min, $kg^{.7}$					
Mouse	25	3.7	0.94	0.10	–
Rat	26.4	5.2	1.02	0.12	13
Monkey	–	2.3	–	0.22	5
Dog	–	82	–	2.8	32

Table 3. Tissue- to blood distribution coefficients for parent PCBs (R) and metabolites (R'). Mouse from reference (13), rat from reference (12), monkey and dog from reference (30)

	4-MCB				44'-DCB				22'455'-PCB				245-HCB				236-HCB			
	Mouse	Rat	Monkey	Dog	Mouse	Rat	Monkey	Dog	Mouse	Rat	Monkey	Dog	Mouse	Rat	Monkey	Dog	Mouse	Rat	Monkey	Dog
Parent, R																				
Muscle	1	1	–	–	2	2	5	4	5	1	–	–	5	4	7	6	–	–	4	4
Skin	10	10	–	–	10	10	50	12	20	7	–	–	35	30	70	30	–	–	40	8
Fat	30	30	–	–	70	70	300	40	200	70	–	–	300	400	500	300	–	–	250	30
Liver	1	1	–	–	5	3	20	6	14	12	–	–	10	12	30	10	–	–	20	2
Metabolite, R'																				
Muscle	0.14	0.14	–	–	0.4	0.4			0.1	0.1	–	–	3	0.3	1	0.2	–	–	0.1	0.1
Skin	0.25	0.25	–	–	0.8	0.3			0.1	0.1	–	–	5	2	3	0.7	–	–	0.5	0.2
Fat	0.4	0.4	–	–	1	0.6			0.4	0.4	–	–	1	2	9	2	–	–	1	0.25
Liver	2	2	–	–	4	5			2	2	–	–	10	4	5	10	–	–	5	10

4 Metabolism

Model-derived parameters for metabolism of the various PCB congeners in several animal species are listed in Table 2. Examination of these values gives some interesting insight into the interspecies variations in metabolic rates for PCB congeners and isomers. The model utilized a first-order rate constant in the liver compartment to characterize the metabolism of PCBs to a single metabolic species. There was no evidence of nonlinearities in metabolism of 22'455'-pentachlorobiphenyl in the rat at 0.6 mg/kg and 6.0 mg/kg (20). The assumption of a single metabolic species simplified the analysis, and was warranted since arene oxide formation is believed to be the rate limiting step in the metabolism by hepatic mixed function oxidases. A further explanation of this assumption is given in Lutz et al. (12). The simplified metabolic scheme utilized by the model should still provide a reasonable representation for the disappearance rate of parent PCB, even if all possible individual metabolites are not specifically identified.

Table 2 suggests that the rates of metabolism for PCBs show some trend with the degree of chlorination of the biphenyl ring. Consider, for example, the metabolism rate constants (metabolic clearance), km, for the rat since the most extensive set of data exists for this species. The km values decrease monotonically from the 4-monochloro to the 22'44'55'-hexachlorobiphenyl. This same trend is true for mouse, dog and monkey. However, chlorine position of the biphenyl ring is also an important determinant of metabolism rate. This fact is suggested by comparison of the km values for the two hexachlorobiphenyl isomers listed in Table 2. The 22'33'66'-HCB shows a metabolism rate that is two orders of magnitude greater than the 22'44'55'-HCB in the rat. An order of magnitude difference between these hexachlorobiphenyls is also noted for the monkey and dog. Kato et al. (21) discussed the metabolism of hexachlorobiphenyl isomers in the rat. They indicate that the rate of metabolism and subsequent excretion depends on the chlorine position rather than simply the degree of chlorination of the biphenyl ring. The meta and para positions appear to be the preferred sites for arene oxide formation which may explain why the km for the 22'33'66'-HCB is even larger than that for the 44'-DCB. Overall, our model derived metabolic rate constants in Table 2 support the contention that adjacent unsubstituted carbons on the biphenyl facilitate metabolism. Arene oxides have been implicated as the reactive intermediate for the covalent binding of PCBs to subcellular macromolecules which could be the first step in carcinogenesis (22). Therefore, the more rapidly metabolized compounds may pose a greater threat of toxicity than the more persistent, but slowly metabolized PCBs.

Interspecies correlations of the metabolic rate constants with body weight or surface area are not apparent from the parameters in Table 2. The direct scaling of the metabolism parameter from mouse, rat, dog or monkey to man is not a simple task. This scaling problem for metabolism constants has been observed and previously reported in the pharmacokinetic modeling literature for other chemicals (23, 24), particularly those that undergo enzymatic reactions *in vivo*. Table 2 shows that km increases monotonically from mouse to rat, monkey and dog corresponding to an increase in body size. When, however, these parameters are normalized by either body weight or body surface area, as presented in Table 2, no consistent pattern of km emerges, and no correlation among these species is evident. The mouse and rat have about the same rate of metabolism per unit of body weight for the PCBs listed in Table 2. On a body weight basis, the metabolic rate constant for the monkey is a bit lower than that of the rat or mouse. However, on a surface area basis, they are comparable. On any basis, the dog metabolic rate constants are significantly greater than those of mouse, rat, or monkey. It is not evident from these modeling studies alone which, if any, of these species would predict PCB metabolism in man. A priori predictions for man based on animal studies would be suspect without additional information about metabolism. Though the present data may not permit one to generalize about predictions of *in vivo* metabolism rate of PCBs among various species, one method that shows considerable potential for overcoming the interspecies limitation is *in vitro-in vivo* correlation. This method is designed to use data on enzyme activities from *in vitro* preparations from individual species such as tissue homogenates, cell suspensions, microsomal preparations, or isolated perfused organs. The enzyme activities determined in this manner, when converted to the proper basis for *in vivo* comparison, can be incorporated into the pharmacokinetic model as a metabolism rate constant. This approach may provide a more accurate method for predicting metabolism rates for that species. Such a procedure has already been reported successfully in the literature for other compounds. For example, Dedrick et al. (24) have discussed the large variability of deaminase activity in various tissues in mouse, monkey, dog and man. They used the Vmax and Michaelis constants determined *in vitro* for the deamination of cytosine arabinoside (Ara-C) and with the aid of a pharmacokinetic model were able to predict tissue and plasma or serum concentrations of Ara-C and its metabolic product (Ara-U) in four species including man (23). Collins et al. (25) used *in vitro* liver homogenate reaction data to predict clearances and half-lives of phenytoin in pregnant and non-pregnant rats. Lin et al. (26) used in vitro data from rat and rabbit liver microsomes to estimate Vmax and Michaelis constants for the deethylation of ethoxybenzamide and incorporated these parameters into a pharmacokinetic model to simulate plasma and tissue concentrations of ethoxybenzamide.

Schnellman et al. (27) have reported results of *in vitro* metabolism studies of several PCB congeners and isomers by hepatic microsomal prepartions from several experimental animals and humans (28). Their data are in general agreement with the results of the pharmacokinetic studies with mouse, rat, monkey and dog in terms of the relative rates of metabolism of a particular PCB compound. For example, Schnellman's data (27) indicated that the dog had the highest values of Vmax (pmoles/nmole P-450/min) for the 44'-DCB and 22'33'66'-HCB compared to rat, monkey, and even human. The dog was the only species that showed any measurable *in vitro* activity for the 22'44'55'-HCB. In our analysis of in vivo experiments, the dog had the fastest metabolism of all PCBs and the highest rate of elimination. According to Schnellman, the rat and monkey had comparable Vmax values for 44'-DCB and 22'33'66'-HCB *in vitro*, and these were similar to the Vmax estimates from human liver microsomal preparations from biopsy specimens. This type of *in vitro* metabolism information can be incorporated into a pharmacokinetic model to make predictions of metabolism rates and disposition of PCBs in humans and should be the subject of future modeling studies. The *in vitro* data must be expressed in a proper form consistent with the units in the model equations. This may not be as simple as it first appears. Parry et al. (29) have discussed the difficulties of interpreting Vmax and KM values from cytochrome P-450 enzyme assays because of the biphasic nature of the lipid bilayer membranes of the microsomal preparations and the aqueous environment of the suspending medium. Ligand-receptor interactions within the membrane involve partitioning which affects the calculation of the dissociation and kinetic constants. However, in principle, the *in vitro-in vivo* correlations can be accommodated within the context of the pharmacokinetic model, but proper care must be excercised. This would allow a more rational method for extrapolating animal metabolism data to humans and would permit predictions of PCB pharmacokinetics in man.

5 Equilibrium Partitioning – Parent and Metabolite

The tissue-to-blood distribution coefficients for parent (R) and metabolite (R′) are listed in Table 3 for a number of PCB congeners that have been modeled in mouse, rat, dog, and monkey. Several general observations can be made regarding these parameters. The rank order of partitioning coefficient for parent compound in each species is fat > skin > liver > muscle > blood. This trend is evident from Figure 6 for the rat and is also illustrated in Figure 8 a, b, and c which shows the concentration of three parent PCBs in the tissue of the monkey as a function of time (30). Note that the concentration ranges can span nearly three orders of magnitude. These results are essentially consistent with the findings in other animal studies (eg. 31–33).

The fat compartment has substantially higher tissue-to-blood partitioning coefficients for parent compound for all the congeners. Of the various congeners, the 245-HCB has the highest fat-to-blood ratio for each animal species. Fat accounts for a disproportionate share of the body burden of the PCBs relative to its fraction of body weight (34, 35). The transients during the distribution phase are often dominated by compartments with large capacity for a chemical but with low blood perfusion rates. Such is the case for PCBs in the fat compartment, and to a lesser extent, the skin compartment. The large distribution coefficient for parent PCB in adipose tissue is not surprising considering the lipophilic nature of PCBs. The distribution in various fat depots varies only slightly. For example, cardiac, perirenal, and subcutaneous fat showed nearly equal concentrations of PCB in animal experiments, although concentration in peritesticular fat was slightly less, perhaps due to lower perfusion rates (36).

A good example of how the fat compartment can exert a significant influence on PCB distribution is illustrated in Figure 7. Here we see data for 22′44′55′-HCB in the blood and fat of rats out to 42 days after a single bolus injection of 0.6 mg/kg (12). The dotted lines represent model simulations that used compartments of constant volume. These simulations overpredict the long-term data. These dis-

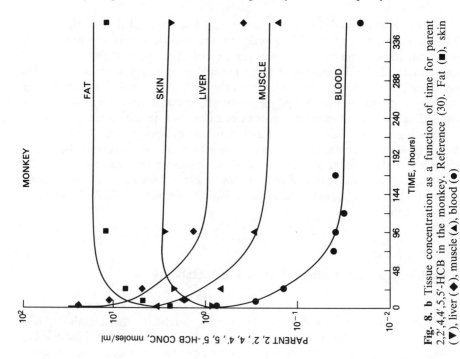

Fig. 8. b Tissue concentration as a function of time for parent 2,2′,4,4′,5,5′-HCB in the monkey. Reference (30). Fat (■), skin (▼), liver (◆), muscle (▲), blood (●)

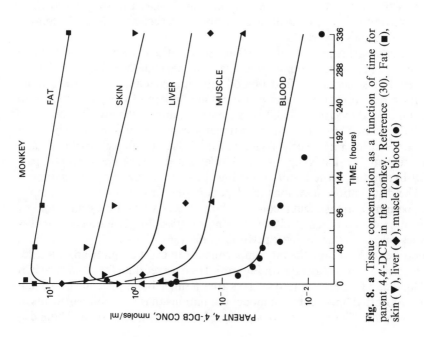

Fig. 8. a Tissue concentration as a function of time for parent 4,4′-DCB in the monkey. Reference (30). Fat (■), skin (▼), liver (◆), muscle (▲), blood (●)

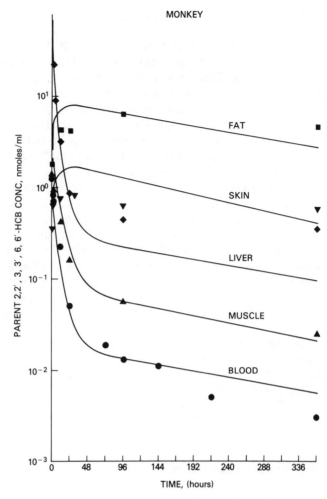

Fig. 8. c Tissue concentration as a function time for parent 2,2′,3,3′,6,6′-HCB in the monkey. Reference (30). Fat (■), skin (▼), liver (◆), muscle (▲), blood (●)

tribution experiments utilized young growing rats at the time of the initial HCB injection, and over the course of 42 days of the experiment, the rats were growing nearly linearly with time from about 250 grams to nearly 400 grams. Also, the adipose content was increasing from seven percent to eleven percent in an asymptotic fashion. This volume change was described by an analytical function of time (12) and incorporated into the mathematical model giving the simulation represented by the solid line in Figure 7. The long-term behavior of this HCB is much better represented by the revised model with changing volume. What the data appeared to depict as washout or elimination of HCB was actually a dilution effect due to the increasing fat volume. Hansen and Welborn (34) have described a similar phenomenon in young swine. This example helps to highlight the utility of pharmacokinetic models as a tool for testing a potential hypothesis of a distribution mechanism.

In general, the tissue-to-blood ratios for PCB metabolites are significantly lower than the corresponding tissue-to-blood ratios for the parent, as seen by the model parameter R' in Table 3. The metabolites, often as a glucuronide conjugate of the parent compound, are less lipophilic and more water soluble than the parent compound. The largest values of tissue-to-blood ratios for metabolite are found in the liver for each species that we studied, which probably relates to the fact that metabolism of PCBs occurs primarily in the liver. Muscle to blood ratios are usually less than 1.0. The overall distribution "space" for PCB metabolites is much smaller than the parent PCB "space".

Specific covalent binding of metabolites to tissue elements has not been incorporated into this pharmacokinetic model of PCB. This shortcoming should be addressed in future model studies since Morales and Matthews (37) have shown that rapidly metabolized PCBs can bind covalently to macromolecular components of tissue such as proteins, DNA, and RNA. Covalent binding of PCB metabolites usually accounts for only a small fraction of the administered dose. However, this fraction may be important biologically since it may adversely affect cellular function and thereby be responsible for certain toxic responses of PCBs. Kinetically, the bound PCB metabolites probably follow a time course consistent with the macromolecule turnover rate (37) which is longer than the clearance values derived from these model simulations. A low concentration "tail" on the data at long time is a common experimental observation that is not well simulated by the model without provision for covalent binding. King et al. (38) have modeled the binding of cisplatin to macromolecules using protein turnover rates as a determinant of platinum kinetics.

6 Excretion-Elimination

The major routes of elimination of PCB are fecal and urinary. In the model, it was assumed that only the metabolized form of the PCBs is excreted by the bile and the urine. This generalization is supported, for the most part, by experimental data (13–17) which showed negligible amounts of parent PCB in feces and urine.

There are some interesting exceptions to this statment. The monkey, for example, showed the presence of small amounts of 44'-DCB, 22'44'55'-HCB, and 22'33'66'-HCB parent compound in the urine. Also, there was evidence of small amounts of 44'-DCB and 22'44'55'-HCB parent compound in monkey's feces, but only at early times after the intravenous injection. The earliest feces sample collected from the monkey was at 24 hours, and it contained approximately 50% parent 22'44'55'-HCB, but samples at later times were generally less than 10% parent. Lutz et al. (30) reported that the composition of bile collected from monkeys was 56% parent 22'44'55'-HCB at 45 minutes, but only 14% parent at 120 minutes. Kato et al. (21) reported the presence of a high percentage of parent

$22'44'55'$-HCB in the feces of rats. In all of these cases, the absolute amount of parent PCB eliminated via the urine or the feces was a small fraction of the total dose, usually less than 5%. Therefore, it seemed reasonable to simplify the model by assuming that only metabolite was excreted in bile and urine.

Aside from the minor amounts of parent PCBs excreted by the bile at early times, PCB may enter the gastrointestinal tract by back diffusion from the gut tissue into the lumenal contents. After an iv bolus injection of a slowly metabolized PCB, high concentrations of PCB appear in all rapidly perfused tissues including the gut. The highly lipophilic PCBs may diffuse across the concentration gradient to the gut lumen contents where it is carried out with the fecal material. As the parent PCB redistributes to other depots such as adipose tissue, or as it is metabolized to larger molecular weight hydrophilic material (eg. glucuronide), the concentration driving force into gut contents diminishes and this mechanism of entry into feces is reduced. Yoshimura and Yamamoto (39) proposed such a mechanism of gut wall transport of $2344'$-tetrachlorobiphenyl in rats since they found only parent compound in the feces of bile-ligated animals. Bungay et al. (40) modeled such a mechanism based on data for kepone transport in the rat intestines. This mechanism of elimination was not incorporated into our pharmacokinetic model of PCB since it represented such a small amount of administered dose. However, we now realize that for the $22'44'55'$-HCB parent compound, it may represent a sizable fraction of the radioactivity collected in the feces (mouse, rat, monkey) since $22'44'55'$-HCB is so slowly metabolized. By assuming that the radioactivity in fecal excretion was only metabolite, our model may overestimate the metabolic clearance, km.

The clearance parameters for biliary and urinary elimination are listed in Table 4. The parameters do not exhibit any clear cut interspecies scaling correlations.

The experimental data for fecal and urinary collections are plotted in Figures 9 a–d for several PCBs in mouse, rat, dog and monkey (12–17). A number of observations can be made. For the higher chlorinated PCBs such as the penta- and the hexachlorobiphenyls, the predominant route of elimination is via the feces for

Table 4. Kidney clearance, k_K, and biliary clearance, k_B, for selected PCBs in several species. References (12, 13, 30)

ml/min	4-MCB		44'-DCB		22'455'-PCB		245-HCB		236-HCB	
	k_B	k_K	k_B	k_K	k_B	k_K	k_B	k_K	k_B	k_K
Mouse (38 g)	0.05	0.05	0.15	0.069	0.10	0.009	0.074	0.018	–	–
Rat 250 g)	0.2	0.2	0.35	0.133	0.3	0.033	0.30	0.03	1.0	0.03
Monkey (5 kg)	–	–	0.083	1.5	–	–	0.70	0.041	0.5	0.4
Dog (12 kg)	–	–	10.2	2.7	–	–	1.8	0.15	7.0	2.0

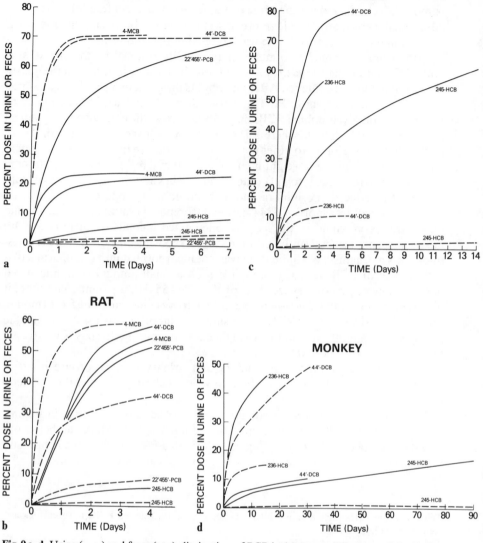

Fig. 9 a–d. Urine (– – –) and feces (——) elimination of PCB in **(a)** mouse (13), **(b)** rat (12), **(c)** dog (15, 16,17), **(d)** monkey (15, 16, 17) as percent of dose versus time after bolus injection of 0.6 mg/ kg

all species listed. For lowest degree of chlorination, the situation appears to reverse. Although the 4-monochlorobiphenyl was studied only in the mouse and rat, urinary excretion from both of these species exceeded fecal excretion. The results for the 44′-dichlorobiphenyl elimination are mixed. The mouse and monkey excrete greater fractions of the dichlorobiphenyl in the urine, whereas the dog and rat excrete DCB mostly in the feces. Of all the species, the dog excreted the greatest fraction of administered dose for the 245-HCB, 90% or more. This excretion reflects the dog's greater metabolic capacity for 245-HCB since metabolism is an important determinant of eventual elimination.

With regard to the preferential fecal elimination of the higher chlorinated PCBs, Smith (17) suggests that a transitional molecular weight exists above which the preferred route of excretion is in the bile, and this transitional molecular weight varies from species to species. For example, the values are 325 for the rat, 475 for the rabbit, and 400 for the guinea pig. A different cutoff molecular weight for different species could result in the type of excretion data illustrated in Figure 9, wherein the middle molecular weight congener, 44'-DCB, varies between urine and feces as to which route of elimination will predominate, depending on the species.

This chapter has reviewed the findings of physiologic pharmacokinetic model analysis of the distribution and disposition of some polychlorinated biphenyl congeners in several animal species. A flow-limited model is adequate for simulating data in most tissues. Analysis of the various pharmacokinetic parameters gives insight into congener-to-congener and species-to-species comparisons. For example, the distribution coefficient for parent compound of all congeners is greatest in the fat tissue of all species studied, with the remaining order generally being skin > liver > muscle > blood. Rates of metabolism vary considerably among congeners, with those congeners having adjacent unsubstituted carbons in the meta and para positions giving the greatest rates. On a body weight basis or a surface area basis, the metabolic clearances show no apparent correlation among mouse, rat, dog, and monkey. *In vitro-in vivo* correlations hold the best promise for interspecies extrapolation of metabolic rates. Clearance occurs primarily in the feces and also in the urine. PCBs are eliminated predominantly as metabolites and their glucuronide conjugates. Physiologic pharmacokinetic models are useful didactic tools for assessing the modes of distribution of PCBs and other chemicals.

7 References

1. Hutzinger O, Choudry GG, Chittim BG, Johnston LE (1985) Formation of polychlorinated dibenzofurans and dioxins during combustion, electrical equipment fires and PCB incineration. In: Environmental Health Perspectives *60*:3
2. McConnell EE (1985) Comparative toxicity of PCBs and related compounds in various species of animals. In: Environmental Health Perspectives *60*:29
3. Allen JR, Carstens LA, Abrahamson LJ (1976) Response of rats exposed to polychlorinated biphenyl for fifty-two weeks. I. Comparison of tissue levels of PCB and biological changes. In: Arch. Environ. Contam. Toxicol. *4*:404
4. Greenberg L, Mayers MR, Smither AR (1939) The systemic effects resulting from exposure to certain chlorinated hydrocarbons. In: J. Ind. Hyg. Toxicol. *21*:29
5. Alvares AP, Fishbein A, Anderson KE, Kappas A (1977) Alterations in drug metabolism in workers exposed to polychlorinated biphenyls. In: Clinical Pharmacol. and Therapeut. *22*:140
6. Fishbein L (1974) Toxicology of chlorinated biphenyls. In: Ann. Rev. Pharmacol. *14*:139
7. Chen HSG, Gross JF (1979) Physiologically based pharmacokinetic models for anticancer drugs. In: Cancer Chemother. Pharmacol. *2*:85

8. Gerlowski LE, Jain RK (1983) Physiologically based pharmacokinetic modeling: Principles and applications. In: J. Pharm. Sci. *72*:1103
9. Dedrick RL (1973) Animal scale up. In: J. Pharmacokin. Biopharm. *1*:435
10. Dedrick RL, Bischoff KB (1980) Species similarities in pharmacokinetics. In: Federation Proc. *39*:54
11. Lutz RJ, Dedrick RL, Zaharko DS (1980) Physiological pharmacokinetics: An *In Vivo* Approach to Membrane Transport. In: Pharmac. Ther. *11*:559
12. Lutz RJ, Dedrick RL, Matthews HB, Eling TE, Anderson MW (1977) A preliminary pharmacokinetic model for several chlorinated biphenyls in the rat. In: Drug Metab. Dispos. *5*:386
13. Tuey DB, Matthews HB (1980) Use of a physiological compartmental model for the rat to describe the pharmacokinetics of several chlorinated biphenyls in the mouse. In: Drug Metab. Dispos. *8*:397
14. Anderson MW, Eling TE, Lutz RJ, Dedrick RL, Matthews HB (1977) The construction of a pharmacokinetic model for the disposition of polychlorinated biphenyls in the rat. In: Clin. Pharmacol. Therapeutic. *22*:765
15. Sipes IG, Slocumb ML, Perry DF, Carter DE (1980) 4,4'-Dichlorobiphenyl: Distribution, metabolism, and excretion in the dog and monkey. In: Toxicol. Appl. Pharmacol. *55*:554
16. Sipes IG, Slocumb ML, Chen HSG, Carter DE (1982) 2362'3'6'-Hexachlorobiphenyl:Distribution, metabolism, and excretion in the dog and the monkey. In: Toxicol. Appl. Pharmacol. *62*:317
17. Sipes IG, Slocumb ML, Perry DF, Carter DE (1982) 2452'4'5'-Hexachlorobiphenyl: Distribution, metabolism, and excretion in the dog and the monkey. Toxicol. Appl. Pharmacol. *65*:264
18. Matthews HB, Anderson MW (1975) Effect of chlorination on the distribution and excretion of polychlorinated biphenyls. In: Drug Metab. Dispos. *3*:371
19. Matthews HB, Tuey DB (1980) The effect of chlorine position on the distribution and excretion of four hexachlorobiphenyl isomers. In: Toxicol. Appl. Pharmacol. *53*:377
20. Matthews HB, Anderson MW (1975) The distribution of 2452' 5'-Pentachlorobiphenyl in the rat. In: Drug Metab. Dispos. *3*:211
21. Kato S, McKinney JD, Matthews HB (1980) Metabolism of symmetrical hexachlorobiphenyl isomers in the rat. In: Toxicol. Appl. Pharmacol. *53*:389
22. Kohli KK, Gupta BN, Albro PW, Mukhtar H, McKinney JD (1979) Biochemical effect of pure isomers of hexachlorobiphenyl (HCB): Fatty livers and cell structure. In: Chem.-Biol. Interact. *25*:139
23. Dedrick RL, Forrester DD, Ho DHW (1972) *In-vitro – in vivo* correlation of drug metabolism – deamination of 1-D-arabinofuranosylcytosine. In: Biochem. Pharmacol. *21*:1
24. Dedrick RL, Forrester DD, Cannon JN, El Dareer SM, Mellett LB (1973) Pharmacokinetics of 1-D-arabinofuranosylcytosine deamination in several species. In: Biochem. Pharmacol. *22*:2405
25. Collins JM, Blake DA, Egner PG (1978) Phenytoin metabolism in the rat-Pharmacokinetic correlation between *in-vitro* hepatic microsomal enzyme activity in *in-vivo* elimination kinetics. In: Drug Metab. Dispos. *6*:251
26. Lin JH, Sugiyama Y, Awazu S, Manabu H (1982) Physiological pharmacokinetics of ethoxybenzamide based on biochemical data obtained *in-vitro* as well as on physiological data. In: J. Pharmacokin. Biopharm. *10*:649
27. Schnellman RG, Vickors FEM, Sipes IG (1985) Metabolism and Disposition of Polychlorinated Biphenyls. In: Hodgson E, Bend J, Philips R (eds) Rev. Biochem. Toxicol. No. 7. Elsevier Press, pp 247–282
28. Schnellman RG, Putnam CW, Sipes IG (1983) Metabolism of 22'33'66'-Hexachlorobiphenyl and 22'44'55'-hexachlorobiphenyl by human hepatic microsomes. In: Biochem. Pharmacol. *32*:3233
29. Parry G, Palmer DN, Williams DJ (1976) Ligand partitioning into membranes: Its significance in determining Km and Ks values for cytochrome P-450 and other membrane bound receptor enzymes. FEBS Letters *67*:123
30. Lutz RJ, Dedrick RL, Tuey D, Sipes IG, Anderson MW, Matthews HB (1984) Comparison of the pharmacokinetics of several polychlorinated biphenyls in mouse, rat, dog, and monkey by means of a physiological pharmacokinetic model. In: Drug Metab. Dispos. *12*:527

31. Muhlebach S, Bickel MH (1981) Pharmacokinetics in rats of 2452'4'5'-hexachlorobiphenyl, an unmetabolized lipophilic model compound. In: Xenobiotica 11:249
32. Matthews HB, Tuey DB (1980) The effect of chlorine position on the distribution and excretion of four hexachlorobiphenyl isomers. In: Toxicol. Appl. Pharmacol. 53:377
33. Morales NM, Tuey DB, Colburn WA, Matthews HB (1979) Pharmacokinetics of multiple oral doses of selected polychlorinated biphenyls in mice. In: Toxicol. Appl. Pharmacol. 48:397
34. Hansen LG, Welborn ME (1977) Distribution, dilution, and elimination of polychlorinated biphenyl analogs in growing swine. In: J. Pharm. Sci. 66:497
35. Wyss PA, Muhlebach S, Bickel MH (1982) Pharmacokinetics of 22'44'55'-hexachlorobiphenyl (6-CB) in rats with decreasing adipose tissue mass. I. Effects of restricting food intake two weeks after administration of 6-CB. In: Drug Metab. Dispos. 10:657
36. Berlin M, Gage J, Holm S (1975) Distribution and metabolism of 2452'5'-Pentachlorophenyl. In: Arch. Environ. Health 30:141
37. Morales NM, Matthews HB (1979) In-vivo binding of 2362'3'6'-Hexachlorobiphenyl and 2452'4'5'-Hexachlorobiphenyl to mouse liver macromolecules. In: Chem.-Biol. Interactions 27:99
38. King FG, Dedrick RL, Farris FF (1986) Physiological pharmacokinetic modeling of cis-Dichlorodiammineplatinum (II) (DDP) in several species. In: J. Pharmacokin. Biopharm. 14:131
39. Yoshimura H, Yamamoto H (1975) A novel route of excretion of 2434'-Tetrachlorobiphenyl in rats. In: Bull. Environ. Contam. Toxicol. 13:681
40. Bungay PM, Dedrick RL, Matthews HB (1981) Enteric transport of chlordecone (kepone) in the rat. In: J. Pharmacokin. Biopharm. 9:309
41. Smith RL (1973) The excretory function of bile. Chapman and Hall, London, p 35

PCBs and Human Health

S. Safe [1]

Humans have been exposed to polychlorinated biphenyl (PCB) mixtures via 3 major pathways. Several thousand workers engaged in the manufacture and application of PCBs have been exposed to relatively high levels of these compounds and serum levels up to 3000 ppb have been measured in occupationally-exposed individuals. The accidental leakage of a PCB-containing industrial fluid into rice oil resulted in the exposure of several thousand individuals in two separate incidents in Japan (1968, Yusho poisoning) and Taiwan (1979, Yu Cheng poisoning). The Yusho/Yu Cheng poisoning victims constitute a second high PCB exposure group. Population surveys have also been shown that most humans are exposed to relatively small concentrations of PCBs through the food chain and constitute the low level exposure group. Recent studies have shown that the relative severity of the toxic symptoms observed in Yusho and Yu Cheng incidents compared to the moderate effects observed in occupationally-exposed workers is primarily due to the high levels of polychlorinated dibenzofurans (PCDFs) identified in the PCB fluid which contaminated the rice oil in the Yusho and Yu Cheng victims. Occupational exposure to PCBs can result in dermal toxicity, hepatic dysfunction and decrease in pulmonary function; however, these effects appear to be reversible after exposure to PCBs is terminated. Based on the severity and duration of the toxic symptoms observed in workers exposed to high levels of PCBs, it is unlikely that environmental exposure to these chemicals leads to any adverse human effects.

1 Introduction

PCBs have been manufactured and marketed for diverse applications since the 1930's; some of these applications include their use as dielectric fluids, heat transfer fluids, organic diluents, plasticizers, flame retardants. PCBs are synthesized by the direct chlorination of biphenyl and the commercial products are mar-

---

[1] Texas A&M University, College of Veterinary Medicine, Dep. of Veterinary Physiology and Pharmacology, College Station, TX 77843-4466, USA

keted according to their chlorine content (1, 2). The unusual versatility of commercial PCBs is associated with the physical and chemical properties of the different formulations and these include resistance to acids and bases, resistance to oxidative breakdown, miscibility with organic solvents, flame retardant characteristics and excellent properties as a dielectric fluid. Individuals who were engaged in the production and/or manufacture of PCB-containing products constitute one of the major groups exposed to relatively high levels of this industrial chemical. Any assessment of adverse human health effects to these compounds should consider the following factors which may influence the potential toxicity of these compounds, namely: (i) the route(s) of exposure, (ii) duration of exposure, (iii) the composition of the commercial PCB products (i.e., degree of chlorination), and (iv) the presence and levels of potentially toxic polychlorinated dibenzofurans. Due to these variables, it would not be surprising to observe significant differences in the effects of PCBs on different groups of occupationally-exposed workers.

The accidental leakage of a PCB-containing heat transfer fluid into rice oil resulted in two serious poisonings in Japan (Yusho poisoning, 1966–1968) and in Taiwan (Yu-Cheng poisoning, 1978–1979) (3–8). Many of the victims of these accidents received a relatively high oral dose of PCBs over a limited time period (weeks-months). Most of the recognized adverse human health effects of PCBs have been determined from studies on the two high level exposure groups, i.e. Yusho/Yu-Cheng victims and exposed workers. This chapter will review the observed toxic effects of PCBs on both groups and attempt to rationalize the data in terms of the differences in the route and duration of exposure and the composition of the commercial products.

2 Human Health Effects of PCBs – Occupational Exposure

Several studies have reported relatively high levels of PCBs in the serum or adipose tissues of individuals occupationally exposed to this commercial product (9–22) and Table 1 summarizes the analytical results from some of these studies. The data clearly demonstrate that occupational exposure to PCBs can result in significant uptake of PCBs as reflected serum levels of these compounds. It was also apparent that the serum concentrations of PCBs were highly variable with some workers exhibiting serum levels in excess of 3,000 ppb (14). Not surprisingly, after exposures to these chemicals were terminated, several reports (15, 19, 20) showed that serum levels tended to decrease. It was also noted that in a group of occupationally exposed women, there was an excellent linear correlation between their serum PCB levels and concentrations in their milk. Lactation is known to effectively decrease overall body burdens of PCBs in women (20, 21); however, it is clear that this may also represent a potential adverse health effect for children suckling mothers occupationally-exposed to this industrial chemical.

The adverse human health effects of occupational exposure to PCBs are summarized in Table 2. Chloracne and related skin problems have been observed in

Table 1. Serum PCB concentrations of occupationally-exposed workers (18)

Facility	No. of suvjects	Arithm. mean (ppb)	Geometric mean (ppb)	Range (ppb)	Refer- ence
Railway car maintenance	86	33.4	–	10–312	(9)
Capacitor plant	34	294	–	0–1700	(10)
Capacitor plants	290	124[a], 48[b]	67[a], 21[b]	6–2530[a], 1–546[b]	(11, 12)
Capacitor plants	80	342[a]	–	41–1319	(13)
Capacitor plant	221	–	119[a], 25.3[b]	1–3330[a], 1–250[b]	(14)
Public utility	14	–	24[a], 24[b]	5–52[a], 7–24[b]	(14)
Private utility	25	–	22[a], 29[b]	9–48[a], 7–250[b]	(14)
Paint manuf.	7	–	64.3	12–190	(15)
Transformer repair	54	–	9.7	1–300	(16)
Capacitor plants (1976)	194	–	363[a], 30[b]	57–2270[a], 6–142[b]	(19)
(1979)	194	–	68[a], 19[b]	12–392[a], 4–108[b]	
Capacitor plant (1973)	20	–	28.8	–	(20)
(1973)	4	–	121.9	–	(20)
(1973)	10	–	68.2	–	(20)
(1980)	20	–	11.2	–	(20)
(1980)	4	–	88.3	–	(20)
(1980)	10	–	67.7	–	(20)

[a] Lower chlorinated biphenyls
[b] Higher chlorinated biphenyls

Table 2. Occupational exposure to PCBs – adverse human health effects

Adverse health effect	Reference
Chlorance and related dermal toxicity	(10, 11, 13, 23–6)
Hepatic dysfunction, including hepatomegaly, increase in serum enzymes (e.g., SGOT, GGTP) and serum lipids, increased drug clearance and induction of microsomal monooxygenases	(9, 11, 13, 14, 16, 19)
Decreases in some pulmonary functions	(26)
Mortality and cancer – no significant increases	(27, 28)

several groups of workers (10, 11, 13, 20, 23–26) and it was suggested that the air concentrations of commercial PCBs > 0.2 mg/m^3 were associated with this effect (26). Since the chloracnegenic effects of PCBs are dependent not only on levels of these toxins in the workplace but also on individual uptake and duration of exposure, it is not possible to accurately determine no-effect exposure levels for dermal toxicity. Moreover in many occupational situations, workers were simultaneously exposed to both PCBs and chlorinated benzenes. It was also reported that after occupational exposure to PCBs was terminated, there was a gradual decrease in the severity and number of dermatological problems in the exposed workers and this paralleled a decrease in their serum levels of PCBs (20). Examination of a group of children nursed by mothers occupationally exposed to PCBs did not show any consistent dermatological symptoms (20).

The effects of occupational PCB exposure to PCBs on the levels of several serum clinical chemical and hematological parameters have been reported by several groups (9, 11, 13, 14, 16, 19, 23). Mildly elevated SGOT and γ-glutamyl transpeptidase (GGTP) suggest some liver damage and induction of hepatic monooxygenase enzymes. These results are not surprising in light of animal studies which report that PCBs frequently cause hepatomegaly accompanied by monooxygenase enzyme induction (24). It was also noted in one study that as PCB serum levels decreased over time serum GGTP levels decreased to normal values. Alvares and coworkers (25) have reported that although the serum enzymes were not elevated in 5 workers exposed to PCBs their antipyrine half-lives were decreased and this was consistent with induced hepatic monooxygenases. Warshaw and colleagues reported a relatively high incidence of pulmonary dysfunction in capacitor workers (26) and the symptoms included coughing (13.8%), wheezing (3.4%), tightness in the chest (10.1%), and upper respiratory or eye irritation (48.2%). The pulmonary toxicity of PCBs in laboratory animals has not been widely reported (24).

Retrospective mortality studies (27) in 2,567 workers (>3 months employment) from two capacitor manufacturing plants gave the following results: mortality in both plants was lower than expected; there were no significant increases in either liver or rectal cancer. An unpublished update of the mortality study (28) in which 7 additional years had elapsed (and therefore there were more deaths in the exposed group) did not alter the initial findings. A study on 31 workers exposed to Aroclor 1254 and employed in a New Jersey petrochemical plant did show an increased incidence of malignant melanomas (0.04% expected; 0.13% observed) and it is evident that more comprehensive long term epidemiologic studies will be required to fully assess the cancer-causing potential of occupational exposure to PCBs. However, it is apparent from most reports that workplace exposure to relatively high levels of PCBs results in limited and moderate toxicity in humans. These toxic symptoms appear to be reversible after exposure to PCBs is terminated and this is accompanied by a decline in serum levels of PCBs.

3 Yusho/Yu-Cheng Poisoning

In 1968 a mass food poisoning was reported in the Fukuoka and Nagasaki prefectures in southwestern Japan. Approximately 1600 individuals suffered a broad spectrum of toxic effects (see Table 3) after consuming rice oil contaminated with a commercial PCB industrial fluid, Kanechlor 400 (3–6). The 1979 Yu-Cheng poisoning of >1900 individuals in Taichung and Changhwa in central Taiwan was also due to the consumption of PCB-contaminated rice oil (7, 8). The most characteristic initial symptom of this toxicosis was chloracne and related dermal problems; in addition, a broad spectrum of effects were observed and these are typified by the toxic symptoms observed by Kuratsune and coworkers (2) (Table 3). Moreover, many of these same symptoms were also observed in Yu-Cheng pa-

Table 3. Percent distribution of signs and symptoms of Yusho and Yu-Cheng Poisoning (28, 29)[a]

Symptoms	Males (Yusho) (n = 89)	Males[a] (Yu Cheng) (n = 15)	Females (Yusho) (n = 100)	Females[a] (Yu Cheng) (n = 12)
Dark brown pigmentation of nails	83.1	86.6	75.0	83.3
Distinctive hair follicles	64.0	40	56.0	41.6
Increased sweating at palms	50.6		55.0	
Acne-like skin eruptions	87.6	86.6	82.0	83.3
Red plaques on limbs	20.2		16.0	
Itching	42.7		52.0	
Pigmentation of skin	75.3		72.0	
Swelling of limbs	20.2		41.0	
Stiffened soles in feet and palm of hands	24.7	46.6	29.0	25
Pigmented mucous membrane	56.2		47.0	
Increased eye discharge	88.8	93.3	83.0	91.6
Hyperemia of conjuctiva	70.8	66.6	71.0	75
Transient visual disturbance	56.2		55.0	
Jaundice	11.2		11.0	
Swelling of upper eyelids	71.9	86.6	74.0	91.6
Feeling of weakness	58.4		52.0	
Numbness in limbs	32.6	53.3	39.0	56
Fever	16.9		19.0	
Learning difficulties	18.0		19.0	
Spasm of limbs	7.9		8.0	
Headache	30.3		39.0	
Vomiting	23.6		28.0	
Diarrhea	19.1		17.0	

[a] Pigmentation of lips, black color of nose, pigmentation of conjuctivae, hypesthesia, deformity of nails, pigmentation of gingivae, amblyopia were also observed

Table 4. Relationship between the amount of Kanechlor-contaminated rice oil consumed and clinical severity of Yusho (29)

Estimated amount of oil consumed (ml)	Nonaffected		Light cases		Severe cases		Total	
	No.	%	No.	%	No.	%	No.	%
<720	10	12	39	49	31	39	80	100
720–1440	0	0	14	31	31	69	45	100
>1440	0	0	3	14	18	86	21	100

tients (28) and it is clear from numerous studies that both poisonings share a common etiology. Kuratsune and coworkers (29) have also demonstrated the dose-response relationship between the consumption of Kanechlor-contaminated rice oil and the severity of Yusho poisoning (Table 4). The severity of the acute poisoning effects of Yusho victims have been monitored since the accident (30, 31) and between 1969–1975 there was significant recovery from the mucocutaneous

Table 5. Distribution of Yusho patients according to skin severity grades in 1971, 1976, and 1981 (31)

Skin severity[a] index	Case numbers		
	1971	1976	1981
0	4	25	56
I	49	27	22
II	32	14	14
III	31	20	13
IV	13	4	2
Total	129	90	107

[a] Grade 0: No skin eruption
Grade I: Circumscribed pigmentation of skin
Grade II: Black comedones
Grade III: Acneiform eruptions
Grade IV: Extensive distribution of the acneiform eruptions

lesions in 64% of the patients. However, it was also reported that other symptoms such as headaches and stomach aches, numbness of the extremities, coughing and bronchial disorders, and joint pains were common in many of these patients. Moreover, children poisoned in the Yusho incident had retarded growth, abnormal tooth development and newborns exhibited systemic pigmentation and were undersized. Table 5 summarizes the changes in the severity of the dermal toxicity of Yusho oil in a group of patients 2, 7 and 12 years after the original outbreak of Yusho poisoning. These qualitative data also illustrate that there is a gradual recovery from the skin problems (30, 31).

It is apparent from these studies that there were major differences in the toxic effects of PCB contaminated rice oil and "normal" industrial PCBs on their respective exposed human populations. The severe acute and chronic effects observed in Yusho victims consuming the contaminated rice oil were not observed in the occupationally-exposed population. Moreover, a close inspection of serum PCB levels in Yusho/Yu-Cheng patients and industrial workers exposed to PCBs were comparable. For example, the mean PCB blood levels of Yu-Cheng victims taken a short time after the accident varied from 39–101.7 ppb (7), whereas the PCB serum levels in occupationally-exposed workers can be much higher (Table 1).

These analytical data suggest that factors other than PCBs must play an important role in the etiology of Yu-Cheng/Yusho poisoning. Several studies have shown that the highly toxic polychlorinated dibenzofurans are present as trace impurities in many commercial Japanese and North American PCB preparations (32–36). The concentrations of PCDFs in these preparations were variable; however, the levels were in the low ppm range. Subsequent work has also identified PCDFs and other chlorinated aromatic hydrocarbons in the PCB industrial fluid which contaminated the rice oil in both the Yusho and Yu-Cheng poisonings (37–42). The PCB, polychlorinated quaterphenyl (PCQ) and PCDF levels in the Japanese and Taiwanese rice oils and some commercial PCB products are summa-

Table 6. Concentrations of PCBs, PCQs and PCDFs and their ratios in various materials (42)

Sample	PCBs	PCQs	PCDFs	PCQs/PCBs	PCDFs/PCBs
Yu-Cheng oil					
Y-1[a]	51	10	0.14	2.0×10^{-1}	2.7×10^{-3}
Y-2[b]	54	18	0.10	3.3×10^{-1}	1.9×10^{-3}
Y-3[b]	69	24	0.18	3.5×10^{-1}	2.6×10^{-3}
Y-4[a]	22	9	$-$[c]	4.1×10^{-1}	$-$
Y-5[b]	113	38	$-$[c]	3.4×10^{-1}	$-$
Average	62	20	0.14	3.3×10^{-1}	2.4×10^{-3}
Yusho oil					
Feb. 5, 1968 (production date)	968	866	7.40	8.9×10^{-1}	7.6×10^{-3}
Feb. 9, 1968 (production date)	151	490	1.90	3.2	1.3×10^{-2}
Feb. 10, 1968 (production date)	155	536	2.25	3.5	1.4×10^{-2}
Average	430	630	3.85	1.5	9.0×10^{-3}
Kanechlor 400	999,800	209	33	2.1×10^{-4}	3.3×10^{-5}

[a, b] Samples collected from a school cafeteria and victims' home, respectively
[c] No analysis

rized in Table 6 (42). The ratio of PCDFs/PCBs in the Yu-Cheng and Yusho oils was 2.4×10^{-3} and 9.0×10^{-3} respectively, whereas the ratio in Kanechlor 400 was 3.3×10^{-5} thus illustrating the relatively high levels of the PCDFs in the toxic contaminated rice oils.

Several recent studies have focused on the identification and quantitation of PCBs and PCDFs in Yusho and Yu-Cheng patients (42–50) and Table 7 (40) summarizes the serum levels of these compounds in several exposed groups and a control Japanese population. It was apparent that the PCB levels in the recently exposed Yu-Cheng patients were higher than the levels in Yusho oil victims, one group of occupationally exposed workers and the control (Japanese population). However, the levels were not significantly greater than those observed in a group of capacitor workers whose last exposure was 9 years prior to the analysis. It was evident from the analytical data that the persistence of the PCDFs in the Yu-Cheng population was the major difference between this group and all the others. The PCDF levels had significantly decreased in the Yusho patients; however, this group had largely recovered from the earlier poisoning. Similar results were observed in another study (Table 8) which also identified the major PCDFs which persisted in the victims of the rice oil poisoning (44). With the exception of the 1,2,4,7,8-pentachlorodibenzofuran congener, all of the PCDFs identified in the adipose tissue and serum were highly toxic congeners and resembled 2,3,7,8-tetra-chlorodibenzo-p-dioxin (TCDD) in their mode of action and toxicity (51–53).

The data noted above suggest that the major etiologic agents in the Yusho/Yu-Cheng poisonings were the PCDFs and the presence of these compounds in the contaminated rice oil account for the marked differences in the toxic symptoms observed in Yusho/Yu-Cheng victims and occupationally-exposed workers.

Table 7. PCB, PCQ, and PCDF levels in the blood of Taiwanese and Japanese poisoned patients, workers occupationally exposed to PCBs, healthy persons, and in toxic rice oils (40, 45)

Sample	No.	Period after termination of exposure (year)	Degree of severity of clinical signs[a]	Concentration (mean ± SD, ppb)			PCBs:PCQs:PCDFs
				PCBs	PCQs	PCDFs	
Taiwanese patients	5	1[b]	None	12 ± 6	1.7 ± 1.1	0.024 ± 0.018	100:14:0.15
	24		Slight	33 ± 13	7.9 ± 3.7	0.062 ± 0.024	100:21:0.16
	14		Moderate	43 ± 11	8.2 ± 3.5	0.079 ± 0.030	100:19:0.19
	24		Heavy	49 ± 20	11.0 ± 5.2	0.100 ± 0.040	100:23:0.20
	67			42 ± 17	8.6 ± 4.8	0.76 ± 0.038	100:20:0.18
Japanese consumers of Yusho oil[h]	56	11		6 ± 4	2.0 ± 2.0	ND[c]	100:32: <0.17
Japanese worker occupationally exposed to fresh PCB[f]	69	9		45 ± 19	ND[d]	–[e]	100: <0.04:–
Japanese worker occupationally exposed to used PCB[g]	3	9		19 ± 11	0.9 ± 0.9	–[e]	100:5:–
Japanese healthy subjects	60			2 ± 1	ND[d]	–[e]	100:0.1:–

[a] Classification according to the report of Goto and Higuchi (1969)
[b] Period after PCB poisoning
[c] <0.01 ppb
[d] <0.02 ppb
[e] Some of blood specimens from workers exposed to PCBs or healthy persons were analyzed, no PCDFs were detected in them at the detection limit of 0.01 ppb
[f] Workers who engaged in charging fresh PCB preparation into condensers in a factory
[g] Workers who engaged in reclaiming PCB used as heat exchanger
[h] Most of them are officially recognized Yusho patients but some are not

Table 8. Concentration of PCBs and PCDF congeners in the tissues of Yu-Cheng infant and the blood of Yu-Cheng and Yusho patients (44)

	Yu-Cheng				Yusho, Blood	
	Baby		Blood			
	Adipose tissue	Liver	Patient A	Patient B	Patient A	Patient B
Total PCBs, ppb	316	27	740	310	4	5
2,3,7,8-Tetrachloro-DF, ppt	17	60	< 30	< 30	<30	<30[a]
1,2,4,7,8-Pentachloro-DF, ppt	14	42	60	40	ND[b]	ND
1,2,3,7,8-Pentachloro-DF, ppt	44	194	30	20	ND	ND
2,3,4,7,8-Pentachloro-DF, ppt	68	91	120	80	3	3
1,2,3,4,7,8-Heptachloro-DF, ppt	88	193	150	60	< 6[b]	< b[a]
Total PCDFs, ppt	231	580	360	200	3	3

[a] Less than detection limit
[b] ND = not detected

Kunita (45) and coworkers have reported the relative toxicity of purified PCBs, PCQs and PCDFs obtained by fractionation of a KC-400 PCB preparation used as a heat exchanger: A reconstituted mixture (Mix-1; PCBs : PCQs : PCDFs; 1 : 1 : 0.01) which approximated the composition of Yusho oil was also prepared and used in the experiment. Dose-response studies clearly showed that only the mixture (Mix 1) and the PCDF fractions were toxic in the rats (i.e., thymic atrophy and body weight loss). A recent study in my laboratory (54) also confirmed that the PCDFs were the major etiologic agents in Yusho oil. Two reconstituted mixtures containing 5 PCDFs (2,3,7,8-tetra-, 1,2,4,7,8-penta-, 1,2,3,7,8-penta-, 2,3,4,7,8-penta, and 1,2,3,4,7,8-hexachlorodibenzofurans;7.4, 6.1, 19.0, 29.4 and 38.1% respectively) and 6 PCBs (2,3′,4,4′,5-penta-, 2,2′,4,4′,5,5′-hexa-, 2,2′,3,4,4′,5′-hexa-, 2,3,3′,4,4′,5-hexa-, 2,2′,3,4,4′,5,5′-hepta- and 2,2′,3,3′,4,4′,5-heptachlorobiphenyls; 5.7, 22.6, 28.2, 12.3, 19.1 and 12.2% respectively) were prepared from pure synthetic standards. These compounds and their relative concentrations were similar to those identified in Yusho victims (15, 40–50). The mixtures were administered in a dose-response fashion to immature male Wistar rats and the biologic and toxic effects measured in this study [i.e., body weight loss, thymic atrophy and the induction of hepatic microsomal aryl hydrocarbon hydroxylase (AHH) and ethoxyresorufin O-deethylase (EROD)] are typically elicited by the toxic PCBs, PCDFs and related halogenated aromatics (28, 37); the results were used to derive the relative potencies of the two mixtures. The PCB dose required for a 4-fold and 40-fold induction of AHH and EROD was 15.0 and 50.1 mg/kg respectively. In contrast, comparable induction by the PCDF mixture was observed at doses of 0.022 and 0.063 mg/kg respectively and confirmed that the PCDFs were 680–790 times more potent (by weight) than the PCBs. The doses of the PCB and PCDF mixtures required to effect a 20% loss of body weight (note: this value was approximately one-half the maximum body weight loss observed in the PCDF treated animals) in the experimental animals (compared to the corn oil control-fed rats) was 398 and 0.44 mg/kg respectively and again illustrates the higher toxicity (>900-fold) of the PCDF mixture. A similar comparison of the ED_{50} (i.e., concentrations required for a 50% reduction in thymus weights in the treated animals compared to the controls) for thymic atrophy shows that PCDFs (ED_{50}, 0.18 mg/kg) were at least 2,210 times more active than the PCB mixture (ED_{50}, 398.1 mg/kg). Since the ratios of PCBs:PCDFs in Yusho/Yu-Cheng victims and in the contaminated rice oils are generally $<500:1$, these results also suggest that the PCDFs are the major etiologic agents in these accidental poisonings.

4 Health Effects of PCBs: Environmental Exposures

The introduction of PCBs into the environment can result in widespread transport of these chemicals and uptake into the food chain. Figure 1 in Chapter 1 illustrates some of the processes which facilitate the distribution of PCBs in the environment and result in their bioaccumulation in humans (55). Serum levels of PCBs in the general population are usually $<10–15$ ppb and these concentrations are significantly lower than observed in industrially-exposed workers (Table 1) (15–22). Since industrial exposure to high levels of PCBs elicits only mild to moderate toxic symptoms which are reversible, it is unlikely that environmental uptake of PCBs results in significant adverse human health effects. Individuals who consume realtively large amounts of fish represent a small subsection of the population exposed to higher levels of PCBs and this is reflected in elevated serum PCB concentrations found in this group (56–58). The mean serum PCB levels in one such group from Triana, Alabama were 17.2 ppb; moreover, there was a correlation between serum PCB concentrations and elevated blood pressure. The incidence of borderline and definite hypertension was increased 30% over the expected values for this population. The authors note that "the colinearity of DDT and PCB serum concentrations in this rural population, exposed to both chemical families through consumption of contaminated fish, precludes any certainty regarding which family of chlorinated hydrocarbons may be correlated with blood pressure." A more recent study reported an inverse correlation between seminal PCB and p,p'-DDE concentrations with sperm motility index in samples with a sperm count less than 20×10^6 cells/ml (59).

Future studies on the adverse human health effects of PCBs will require more standardized methods of PCB analysis and the development of more sensitive techniques to determine subtle toxic effects caused by relatively low exposures (e.g., environmental) to these chemicals. It should be noted that humans are environmentally exposed to a broad range of chemicals which may act through similar, overlapping or different mechanisms. Therefore, the best indicators of the adverse health effects of PCBs and their role in carcinogenicity will be derived from the continuing epidemiological and retrospective studies on occupationally-exposed workers.

5 Acknowledgements

The financial assistance of the Texas Agricultural Experiment Station, the National Institutes of Health, and the Environmental Protection Agency are gratefully acknowledged.

6 References

1. Hutzinger O, Safe S, Zitko V (1974) The chemistry of PCBs. CRC Press, Cleveland, Ohio
2. Brinkman VA, DeKok A (1980) Production, properties and usage. In: Kimbrough RD (ed) Halogenated Biphenyls, Terphenyls, Naphthalenes, Dibenzodioxins and Related Products. Elsevier/North Holland, Amsterdam, New York, Oxford, Chapt. 7, pp 1–40
3. Higuchi K (ed) (1976) PCB poisoning and pollution. Kodansha, Ltd. and Academic Press, Tokyo, London, New York, Chap. 1, p 5
4. Kuratsune M, Yoshimura T, Matsuzaka J, Yamaguchi A (1972) Epidemiological study on Yusho, a poisoning caused by ingestion of rice oil contaminated with a commercial brand of polychlorinated biphenyls. Environ. Health Perspect. 1:119
5. Kuratsune M, Yusho (1980) In: Kimbrough RD (ed) Halogenated Biphenyls, Terphenyls, Naphthalenes, Dibenzodioxins and Related Products. Elsevier/North Holland, Biomedical Press, Amsterdam, p 28
6. Masuda Y, Kuroki H, Yamaryo T, Haraguchi K, Kuratsune M, Hsu ST (1982) Comparison of causal agents in Taiwan and Fukuoka PCB poisonings. Chemosphere 11:199
7. Hsu S-T, Ma C-I, Kwo-Hsiung S, Wu SS, Hsu NH-M, Yeh CC, Wu SB (1985) Discovery and epidemiology of PCB poisoning in Taiwan: a four-year followup. Environ. Health Persp. 59:5
8. Kuratsune M, Shapiro RE (eds) (1984) PCB Poisoning in Japan and Taiwan. Alan R. Liss., New York
9. Chase KH, Wong O, Thomas D, Berney BW, Simon RK (1982) Clinical and metabolic abnormalities associated with occupational exposure to polychlorinated biphenyls (PCBs). J. Occup. Med. 24:109
10. Ouw HK, Simpson GR, Siyali DS (1976) Use and health effects of Aroclor 1242, a polychlorinated biphenyl, in an electrical industry. Arch. Environ. Health 31:189
11. Fischbein A, Wolff MS, Lilis R, Thornton J, Selikoff IJ (1979) Clinical findings among PCB-exposed capacitor manufacturing workers. Ann. N.Y. Acad. Sci. 320:703
12. Wolff MS, Fischbein A, Thornton J, Rice C, Lilis R, Selikoff IJ (1982) Body burden of polychlorinated biphenyls among persons employed in capacitor manufacturing. Int. Arch. Occup. Environ. Helath 49:199
13. Maroni M, Columbi A, Arbosti G, Cantoni S, Foa V (1981) Occupational exposure to polychlorinated biphenyls in electrical workers. II. Health Effects. Brit. J. Ind. Med. 38:55
14. Smith AB, Schloemer J, Lowry LK, Smallwood AW, Ligo RN, Tanaka S, Stringer W, Jones M, Hervin R, Glueck CJ (1982) Metabolic and health consequences of occupational exposure to polychlorinated biphenyls (PCBs). Brit. J. Ind. Med. 39:361
15. TakamatsuM, Oki M, Katsuyoshi M, Inoue Y, Hirayama Y, Yoshizuka S (1985) Survey of workers occupationally exposed to PCBs and of Yusho patients. Environ. Health Perspect. 59:91
16. Emmett EA (1985) Polychlorinated biphenyl exposure and effects in transformer repair workers. Environ. Health Perspect. 60:185
17. Fischbein A (1985) Liver function tests in workers with occupational exposure to polychlorinated biphenyls (PCBs): comparison with Yusho and Yu Cheng. Environ. Health Perspect. 60:145

18. Kreis K (1985) Studies on populations exposed to polychlorinated biphenyls. Environ. Health Perspect. *60*:193
19. Lawton RW, Ross MR, Feingold J, Brown JF (1985) Effects of PCB exposure on biochemical and hematological findings in capacitor workers. Environ. Health Perspect. *60*:165
20. Hara I (1985) Health status and PCBs in blood of workers exposed to PCBs and of their children. Environ. Health Perspect. *59*:85
21. Kuwabara K, Yakushiji T, Watanabe I, Yoshida S, Koyama K, Kunita N, Hara I (1978) Relationship between breast feeding and PCB residues in blood of the children whose mothers were occupationally exposed to PCBs. Occup. Environ. Health *41*:189
22. Wolff MS (1985) Occupational exposure to polychlorinated biphenyls. Environ. Health Perspect. *60*:133
23. Steinberg KK, Freni-Titulaer LWJ, Rogers TN, Burse VW, Mueller PW, Stohr PA, Miller DT (1980) Effects of polychlorinated biphenyls and lipemia on serum analytes. J. Toxicol. Envir. Health *19*:369
24. Safe S (1984) Polychlorinated biphenyls (PCBs) and polybrominated biphenyls (PBBs): biochemistry, toxicology and mechanism of action. CRC Crit. Rev. Toxicol. *13*:319
25. Alvares AP, Fischbein A, Anderson KE, Kappas A (1977) Alterations in drug metabolism in workers exposed to polychlorinated biphenyls. Clin. Pharmacol. Ther. *22*:140
26. Warshaw R, Fischbein A, Thornton J, Miller A, Selikoff IJ (1979) Decrease in vital capacity in PCB-exposed workers in a capacitor manufacturing facility. N.Y. Acad. Sci. Ann. *320*:277
27. Brown DP, Jones M (1981) Mortality and industrial hygiene study of workers exposed to polychlorinated biphenyls. Arch. Environ. Health *36*:120
28. Li WM, Chen CJ, Wong CK (1981) PCB poisoning of 27 cases in three generations of a large family. Clinical Med. (Taipei) 7:23
29. Kuratsune M, Yoshimura T, Matsuzaka J, Yamaguchi A (1972) Epidemiologic study on Yusho, a poisoning caused by ingestion of rice oil contaminated with a commercial brand of polychlorinated biphenyls. Environ. Health Perspect. *1*:119
30. Urabe H, Koda H, Asahi M (1979) Present state of Yusho patients. Ann. N.Y. Acad. Sci. *320*:273
31. Urabe H, Asahi M (1984) Past and current dermatological status of Yusho patients. Amer. J. Ind. Med. 5:5
32. Bowes GW, Mulvihill MJ, Simoneit BRT, Burlingame AL, Risebrough RW (1975) Identification of chlorinated dibenzofurans in American polychlorinated biphenyls. Nature *256*:305
33. Bowes GW, Mulvihill MJ, DeCamp MR, Kende AS (1975) Gas chromatographic characteristics of authentic chlorinated dibenzofurans: identification of two isomers in American and Japanese polychlorinated biphenyls. J. Agr. Food Chem. *23*:1222
34. Morita M, Nakagawa J, Akiyama K, Mimura S, Isono N (1977) Detailed examination of polychlorinated dibenzofurans in PCB preparations and Kanemi Yusho oil. Bull. Environ. Cont. Toxicol. *18*:67
35. Miyata H, Nakamura A, Kahimoto T (1976) Separation of polychlorodibenzofurans (PCDFs) in Japanese commercial PCBs (Kanechlors) and their heated preparation. J. Food Hyg. Soc. Japan *17*:227
36. Miyata H, Kashimoto T (1976) The finding of polychlorodibenzofurans in commercial PCBs (Aroclor, Phenoclor and Clophen) (In Japanese). J. Food Hyg. Soc. Japan *17*:434
37. Miyata H, Kashimoto T, Kunita N (1977) Detection and determination of polychlorodibenzofurans in normal human tissues and Kanemi rice oils caused "Kanemi Yusho" (In Japanese). J. Food Hyg. Soc. Japan *19*:260
38. Buser HR, Rappe C, Gara A (1978) Polychlorinated dibenzofurans (PCDFs) found in Yusho oil and in used Japanese PCB. Chemosphere 7:439
39. Chen PH, Chang KT, Lu YD (1981) Polychlorinated biphenyls and polychlorinated dibenzofurans in the toxic rice-bran oil that caused PCB poisoning in Taichung. Bull. Environ. Contam. Toxicol. *26*:489
40. Kashimoto T, Miyata H, Kunita S, Tung TC, Hsu ST, Chang KJ, Tang SY, Ohi G, Nakagawa J, Yamamoto SI (1981) Role of polychlorinated dibenzofuran in Yusho (PCB poisoning). Arch. Environ. Health *36*:321

41. Chen PH, Lu YD, Yang MH, Chen JS (1981) Toxic compounds in the cooking oil which caused PCB poisoning in Taiwan. II. The presence of polychlorinated quaterphenyls and polychlorinated terphenyls (In Chinese). Clin. Med. (Taipei) 7:77

42. Miyata H, Fukushima S, Kashimoto T, Kunita N (1985) PCBs, PCQs and PCDFs in the tissues of Yusho and Yu-Cheng patients. Environ. Health Perspect. 59:67

43. Kashimoto T, Miyata H, Fukushima S, Kunita N, Ohi G, Tung T-C (1985) PCBs, PCQs and PCDFs in blood of Yusho and Yu-Cheng patients. Environ. Health Perspect. 59:73

44. Masuda Y, Kuroki H, Haraguchi K, Nagayama J (1985) PCB and PCDF congeners in the blood and tissues of Yusho and Yu-Cheng patients. Environ. Health Perspect. 59:53

45. Kunita N, Kashimoto T, Miyata H, Fukushima S, Hall S, Obana H (1984) Causal agents of Yusho. Amer. J. Ind. Med. 5:45

46. Kuroki H, Masuda Y (1978) Determination of polychlorinated dibenzofuran isomers retained in patients with Yusho. Chemosphere 7:771

47. Rappe C, Buser HR, Kuroki H, Masuda Y (1979) Identification of polychlorinated dibenzofurans (PCDFs) retained in patients with Yusho. Chemosphere 8:259

48. Masuda Y, Kuroki H, Yamaryo T, Haraguchi K, Kuratsune M, Hsu ST (1982) Comparison of causal agents in Taiwan and Fukuoka PCB poisonings. Chemosphere 11:199

49. Kashimoto T, Miyata H, Kunita N (1981) The presence of polychlorinated quaterphenyls in the tissues of Yusho victims. Fd. Cosmet. Toxicol. 19:335

50. Nagayama J, Masuda Y, Kuratsune M (1977) Determination of polychlorinated dibenzofurans in tissue of patients with "Yusho". Food Cosmet. Toxicol 15:195

51. Safe S (1986) Comparative toxicology and mechanism of action of polychlorinated dibenzo-p-dioxins and dibenzofurans. Ann. Rev. Pharmacol. Toxicol. 26:371

52. Mason G, Sawyer T, Keys B, Bandiera S, Romkes M, Piskorska-Pliszczynska J, Zmudzka B, Safe S (1985) Polychlorinated dibenzofurans (PCDFs): correlation between in vivo and in vitro structure-activity relationships. Toxicol. 37:1

53. Bandiera S, Farrell K, Mason G, Kelley M, Romkes M, Bannister R, Safe S (1984) Comparative toxicities of the polychlorinated dibenzofuran (PCDF) and biphenyl (PCB) mixtures which persist in Yusho victims. Chemosphere 13:507

54. Bandiera S, Sawyer T, Romkes M, Zmudzka B, Safe L, Mason G, Keys B, Safe S (1984) Polychlorinated dibenzofurans (PCDFs): effects of structure on binding to the 2,3,7,8-TCDD cytosolic receptor protein, AHH induction and toxicity. Toxicol. 32:131

55. Safe S, Safe L, Mullin M (1985) Polychlorinated biphenyls (PCBs)-congener-specific analysis of a commercial mixture and a human milk extract. J. Agric. Food Chem. 33:24

56. Kreiss K, Zack MM, Kimbrough RD, Needham LL, Smrek AL, Jones BT (1981) Association of blood pressure and polychlorinated biphenyl levels. J. Am. Med. Assoc. 245:2505

57. Humphrey HEB (1983) Evaluation of humans exposed to water-borne chemicals of the Great Lakes. Final report for EPA Cooperative Agreement CR807192

58. Baker EL, Landrigan PJ, Glueck CJ, Zack MM, Liddle JA, Burse VW, Housworth WJ, Needham LL (1980) Metabolic consequences of exposure to polychlorianted biphenyls (PCBs) in sewage sludge. Am. J. Epidemiol. 112:553

59. Bush B, Bennett AH, Snow JT (1986) Polychlorobiphenyl congeners, p,p'-DDE and sperm function in humans. Arch. Environ. Contam. Toxicol. 15:333

Subject Index

The Handbook of Environmental Chemistry

Editor: O. Hutzinger

Springer-Verlag
Berlin Heidelberg New York
London Paris Tokyo

Environmental Management

An International Journal for Decision Makers and Scientists

Editor-in-Chief: D. Alexander, Amherst

Encouraging the unusual, **Environmental Management** is a unique journal which presents complementary and contradictory ideas within the format of a single, international publication. The journal is designed to enable environmental scientists, engineers, attorneys, sociologists, and other professionals to pinpoint and properly assess environmental problems and examine interdisciplinary solutions. **Environmental Management** focuses on real problems by providing a forum for the discussion of ideas, findings, and methods that have been, and can be applied to individual environmental management programs.
Covering a broad spectrum of conservation, preservation, reclamation, and utilization, the journal publishes material dealing with ecological modeling, resource management, energy, hazard response, environmental monitoring, and hazardous substances.

Archives of Environmental Contamination and Toxicology

Editor: A. Bevenue, San Mateo
Associate Editor: **M. C. Bowman,** Mount Ida

An international, interdisciplinary journal of full-length articles, **Archives of Environmental Contamination and Toxicology** covers original experimental and theoretical research pertaining to environmental contamination and toxicology. Detailed reports of significant advances and discoveries in the fields of air, water, and soil contamination and pollution are published. In addition, the results of other research in disciplines concerned with the introduction, presence, and effects of deleterious substances in the totel environment are documented in **Archives of Environmental Contamination and Toxicology.**

Bulletin of Environmental Contamination and Toxicology

Editor-in-Chief: H. Nigg, Lake Alfred

Associate Editors: *Analytical Methodology* Y. Iwata, Ricmond, CA; *Aquatic Toxicology* D. R. M. Passino, Ann Arbor; *Environmental Distribution* J. Adams, Silver Spring; *Metabolism and Biochemistry* R.S.Pardino, Reno; *Physiological and Pathological Studies* J. M. King, Ithaca; *Toxicology* J. B. Knaak, Sacramento

Dedicated to the rapid publication of camera-ready contributions, **Bulletin of Environmental Contamination and Toxicology** disseminates advances and discoveries in the areas of air, soil, water, and food contamination and pollution. Descriptions of methods, procedures, and techniques are designed to allow readers to apply them to their own laboratory work. Through articles which are briefer than those found in archival journals, the **Bulletin of Environmental Contamination and Toxicology** provides a meeting ground for researchers to share in new discoveries as soon as they are made.

For further information or sample copies of these journals, please contact the Springer-Verlag office nearest you.

Springer-Verlag
Berlin Heidelberg New York
London Paris Tokyo